W9-AWQ-977

Contents

Introduction

This is the fourth in the series of Butterworth-Heinemann pictorial guides. Like the others - on marketing, selling services and products, and on financial management - this book addresses a serious subject in a very unserious way. We, the authors, hope you will find it as entertaining to read as we did to write.

Total quality management does not, at first sight, seem a subject for a book of this nature, being the territory of recondite theory, mission statements, strictly regulated working practices and other stern considerations. However, on closer examination, the management of organizations is about the way people agree to work together, and is therefore about human nature, human fallibility and human aspiration. So this book is the view of quality management at ground level. It does not seek to replace serious text books, but we hope you, the reader, will find a link to real-life situations at one end of its spectrum and a relatively painless introduction to the theory of total quality management at the other.

In spite of the picture book format this is not a comic book. The medium is well suited to telling a story as well as allowing for the easy integration of graphical illustrations and diagrams, and there are numerous small stories here which illustrate the universal nature of the benefits of TQM and which demonstrate the strong theoretical foundation upon which the idea of total quality management is based.

At the heart of the book is the Oakland TQM model. The model demonstrates that all organizational activities are interactions - links in a chain which ends at the external customer. Examine each link and you will find a transaction between an internal supplier and customer, each one of which is vital to the success of the organization. Each link is defined by a process of delivery and reception between the participants, so the processes are the core activity around which everything else revolves.

To make the supplier-customer chain work smoothly, the core is surrounded by layers of protection and enablement - the management of the processes through systems which are measured and improved through the use of appropriate tools and teamwork. But to make it all function the organization must be able and willing to change to meet its challenge: everyone in it must have the right attitude - it must have the right culture and be sufficiently committed to embark on a policy of constant improvement. This book is dedicated to the task of showing how essential this model is to the well-being of any organization and its employees.

John Oakland and Peter Morris

October 1996

*In business, a major driving force is the element of **competition**. A supplier's competitiveness is determined by the **quality** of its product or service.*

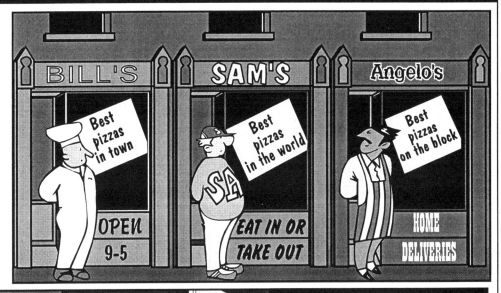

We're going to look at what makes one business more successful than another - what gives a business the reputation for **quality**

Angelo's pizza business is more successful than his competitors. How does he do it?

Angelo pays attention to satisfying his customers. He's constantly watching the reliability, delivery and price of his service.

That was quick!

We deliver in half an hour or you get it free

On the knee of my mama I learned that the price of quality is **constant vigilance**

A company's reputation is based upon the customer's perception of its 'quality'

My uncle found a nail in one in 1983

Hey, what's wrong with my pizzas?

Once a business gets a poor reputation it is very hard to lose it - and it can easily become a national reputation..

Brits can't make pizzas anyway

So what is 'quality'?

What is it that makes people prefer one product or service to others?

The experts and gurus have various definitions...

The totality of features and characteristics of a product or service that bear on its ability to satisfy stated needs

IT'S THIS

The total composite product and service characteristics of marketing, engineering, manufacture and maintenance through which the product and service in use will meet the expectation by the customer

NO IT'S THIS

...but they all add up to the same thing:

Meeting the customer's requirements

But the requirements of each customer - even for the same kind of product - can be completely different...

I bought this watch in the Singapore flea market for $12 about eight years ago. It keeps perfect time, but I must admit it doesn't look much. However, I regard it as excellent value for money. Not so my wife...

I think a watch is a device for telling me the time.

We are **not** going out with that awful tin thing on your wrist!
Get yourself a decent gold watch

She thinks a watch is essentially a piece of jewellery.

The successful business meets all its customers' requirements

The aim of the supplier is to **delight** the customer

Angelo knows how to delight his customers by giving them more than they have asked for.

Defining the customer

The stewardess feels she has let the passengers down..... but someone has let her down

The passengers did not get their bread rolls because there was a break in the chain of supply. The customers were not delighted.

But our guru has a query.

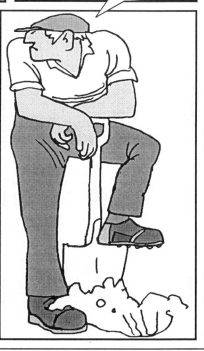

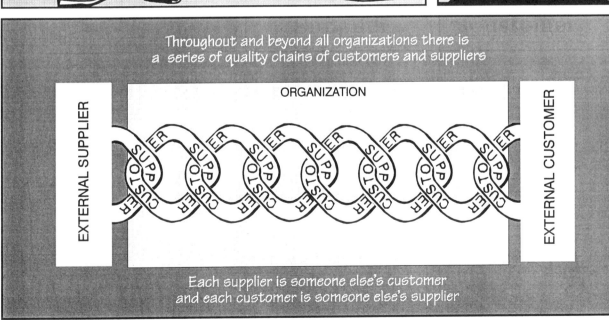

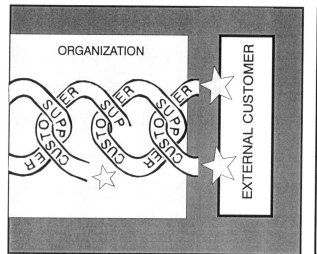

And the consequences of a small break in the internal chain can be proportionally greater when they get to the external customer/supplier interface

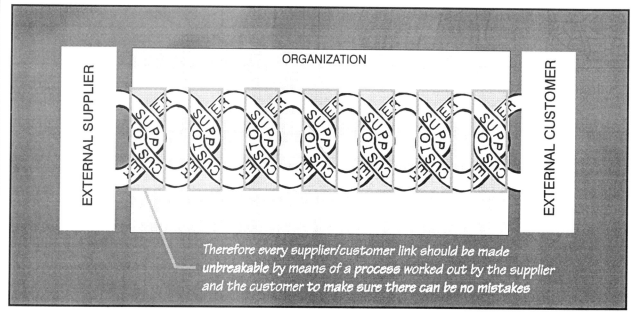

Therefore every supplier/customer link should be made *unbreakable* by means of a process worked out by the supplier and the customer to make sure there can be no mistakes

Quality must be built in from the beginning of an organization's activities, not 'inspected in' at the end

The two-way street

Ensuring quality is a two-way process.

The supplier should meet the customer's requirements. At the same time the customer has a responsibility to make the job of the supplier as easy as possible within the limitations of the supplier-customer relationship...

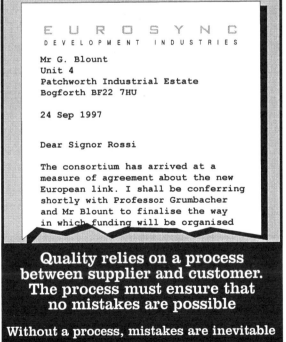

The boss was lucky to get a reasonable letter. Of course, the secretary had to read the proposal herself to be able to write an intelligent reply... Sometimes it's the other way round:

Quality relies on a process between supplier and customer. The process must ensure that no mistakes are possible

Without a process, mistakes are inevitable

Quality will only improve if people want it to

Motivation...

Quality chains must be built by everyone in the organization - even those who do not see the service or product being supplied to the outside customer.

A problem...

Listen, I get here at 8, I work until 4, turning out spindles all day long. I get paid to do my job. It's management's job to manage, they get paid more than I do to do it. You go and ask them how I can build in quality. Don't ask me. I only work here.

The operator has lunch with his friend

What's wrong with your machine, Phil? Half your spindles are out of spec. You're slowing down our assembly line

Ask the supervisor. I'm only the operator

Wrong. **You** are the expert when it comes to that machine. It's up to **me** to tell you what I want and it's up to **you** to try to do it

The key to motivation - and therefore quality - is for each member of the organization to have a well-defined customer - someone identifiable to whom the individual gives an output.

To achieve quality throughout an organization, each person in the quality chain must interrogate each interface as follows:

Customers

- Who are my immediate customers?
- What are their true requirements?
- How do I find out what the requirements are? How can I measure my ability to meet these requirements?
- Do I have the capability to meet the requirements? (What must we change?)
- Do I continually meet these requirements? (If not, why not? What stops me doing my job properly?)
- How do I monitor changes in the requirements?

Suppliers

- Who are my immediate suppliers?
- What are my true requirements?
- How do I communicate my requirements?
- Do my suppliers have the capability to measure and meet my requirements?
- How do I inform them of changes in the requirements?

1 Summary

1   A supplier's competitiveness is determined by the quality of the services or products provided.

2   A reputation for quality is hard won - and easily lost.

3   'Quality' means meeting customers' requirements.

4   Different customers have different requirements of the same product or service. The supplier must determine the customers' requirements and try to exceed them. The supplier should aim to delight the customers.

5   In every organization and beyond there is a series of quality chains of customers and suppliers.

6   As well as external suppliers and customers there are internal 'customers' and 'suppliers' within every organization.

7   At the interface between customer and supplier there must be a process which will guarantee the success of each transaction. Any mistakes will be rectified before reaching the next link in the chain.

8   The process is two-way: it relies on the interaction between supplier and customer.

9   If the quality chain is broken at any point, the effects of the breakdown will be most apparent at the interface with the outside customer.

10 Quality will only improve if people are motivated to make it do so. The key to motivation is to identify a well-defined customer - an individual to whom the supplier gives an output.

11 Quality should be built-in to the organization's activities, not 'inspected-in' at the end.

12 The price of quality is continual re-examination of the requirements of the customer and of the organization's efforts to meet them. This means a policy of continuous improvement.

13 The benefits of quality are increased market share, decreased costs, improved delivery and productivity, and reduced waste.. Achieving this requires **company-wide** quality improvement.

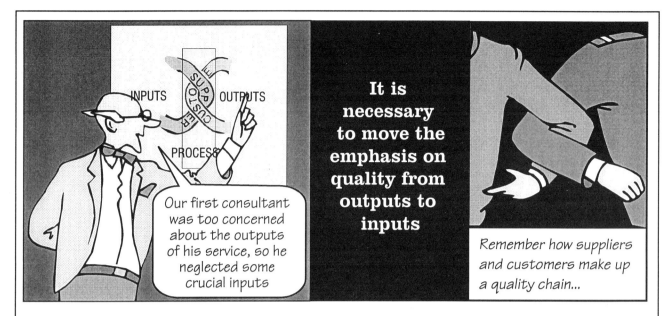

It is necessary to move the emphasis on quality from outputs to inputs

Our first consultant was too concerned about the outputs of his service, so he neglected some crucial inputs

Remember how suppliers and customers make up a quality chain...

...and the processes worked out between them enables a satisfactory exchange to be made between supplier and customer at every link in the quality chain. Any process will be flawed, and be a source of problems, if the supplier does not pay enough attention to his or her inputs

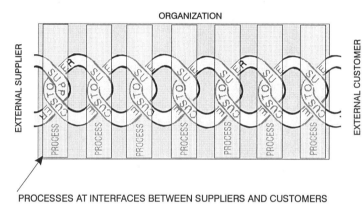

ORGANIZATION

EXTERNAL SUPPLIER

EXTERNAL CUSTOMER

PROCESS

PROCESSES AT INTERFACES BETWEEN SUPPLIERS AND CUSTOMERS

Yes

So when a company or unit inspects its products just before they pass on to their next customer...

...it is, in effect, asking itself: "Are they good enough to sell?" - or: "Will they satisfy the requirements of the next process?" This is not quality control. It is...

No!

Detection!

Quality is being 'controlled' by CHECKING and REJECTING WASTE!
Attempts are being made to 'pour on' quality at the END of the process.

The question being asked is:

Have we done the job correctly?

Whereas what should be asked is:

Are we **capable** of doing the job correctly?

To get the answer 'YES'
a total quality management approach is needed:

satisfactory methods
satisfactory materials
satisfactory equipment
satisfactory skills & knowledge
satisfactory instruction
and a satisfactory

process

What is a process?

It is the transformation of

a set of inputs   into   a set of outputs

What are inputs and outputs?

Inputs are
ACTIONS
METHODS
OPERATIONS

TED-O-MATIC

Outputs are
PRODUCTS
SERVICES

TEDPAK

Inputs are
ACTIONS
METHODS
OPERATIONS

An OUTPUT is
something that
can be transferred
to a customer

Teddy
Teddy

Everything in the work we do is *a process*

As we contribute to our organization's products or services we involve ourselves in processes with the people before us in the chain and with those following us. It's usual for business activities to consist of numerous simultaneous chains - for example there would be a documentation chain at the same time as a production chain for any given product or customer, and this would be multiplied by the number of products and customers. There would be other process chains, such as research and development, administration, maintenance, and so on.

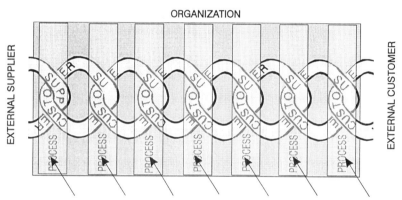

PROCESSES AT INTERFACES BETWEEN SUPPLIERS AND CUSTOMERS

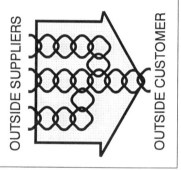

So a more complicated (but still simplified) chain might look like this:

Numerous processes go on simultaneously throughout the organization

Any particular process can be analysed by examining the inputs and outputs. The examination will determine some of the actions necessary to improve quality.

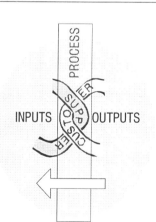

INPUTS OUTPUTS

It is often necessary to move the emphasis in controlling quality from outputs to inputs

In fact, if you define, monitor and control **inputs** in order to meet the requirements of the customer you should produce a satisfactory **output** every time

16

At every supplier-customer interface there is a process so every task throughout an organization is a process.

"By making sure our processes are sound we know we can do the job correctly!"

Every part of every organization, regardless of its size, will have some kind of input to another part of the organization or to an outside customer.

A **process** which will be unique to that situation will result in an output which is at the same time an input to another part of the organization or to the customer

The organization is able to give a positive answer to the first question:

Are we capable of doing the job correctly?

The second question that needs to be asked is:

Do we continue to do the job correctly?

If, from data evidence, **the answer to this is also YES, we must have done the job correctly** and therefore

THERE REALLY IS NO NEED FOR QUALITY

DETECTION

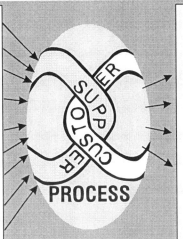

"By continuing to do our job correctly we can guarantee the quality of our product...every time"

QUALITY CONTROL CHART

Stuffing Quantity per bear
+5%
±0%
−5%

The strategy of detection has been replaced by one of prevention

Attention must be concentrated at the beginning of a process, at the inputs, to make sure they are capable of meeting the requirements of the process. This is a managerial responsibility

Materials
Procedures
Methods
Information (including specifications)
People
Skills
Knowledge
Training
Plant/equipment

SUPPLIER
CUSTOMER
PROCESS

Products
Services
Information
Paperwork

Inputs

Outputs

Quality control and quality assurance

I should get it right, because:

I can read my boss's writing;

I know my computer and the word-processor package;

I know the **requirements**

This is quality control

The control of quality can only be carried out at the point of operation

The act of inspection is not quality control.

Did you get it right?

This is NOT quality control

Must make sure...

When the answer to the question "Have we done the job correctly?" is given indirectly by answering "Yes" to the questions "Are we capable of doing the job correctly?" and "Do we continue to do the job correctly?", then we have **assured quality,** and the activity of detecting is replaced by one of **quality assurance,** and the products are the outcome of an effective system.

Keeping correct data of inputs and outputs will audit the system and help prevent problems.

Have we done the job correctly?

Production Charts

Yes!
Yes!
Yes!

Quality assurance will cut across departmental boundaries

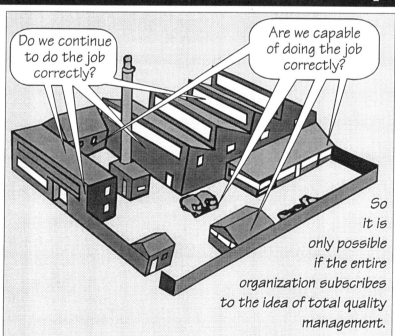

Do we continue to do the job correctly?

Are we capable of doing the job correctly?

So it is only possible if the entire organization subscribes to the idea of total quality management.

Definitions

Quality control

Activities/techniques employed to achieve and maintain the quality of a product, process or service. It monitors; it also finds and helps eliminate causes of problems.

Quality assurance

Prevention of quality problems through planned and systematic activities (including documentation). It includes establishing a good quality management system and an auditing mechanism for the system.

1. Inspecting-out faults at the end of a process is not quality control. It is *detection,* based on asking the question "Have we done the job correctly?"

2. In this case, an attempt is made to 'pour on' quality at the end of a process.

3. Instead of asking the question "Have we done the job correctly?" an organization should ask, first, "Are we capable of doing the job correctly?"

4. To get the answer "Yes" a Total Quality Management approach is needed, satisfactory methods, materials, equipment, knowledge, training, instruction, people with their skills; and a satisfactory process.

5. A process is the transformation of a set of inputs into a set of outputs. An output is something which can be transferred to the customer.

6. Every business interaction within an organization is a process, many of which take place for every product or service offered..

7. A process can be analysed and improved by examining its inputs and outputs.

8. To produce satisfactory outputs it is necessary to define, control and monitor inputs.

9. Nearly all inputs are outputs from other processes often elsewhere in the same organization.

10. The second question that needs to be asked is "Do we continue to do the job correctly?" If the answer to this is also "Yes" the organization must be doing the job correctly, but instead of the task being one of defect or error detection it has become one of prevention. The detection activity is replaced by quality assurance.

11. Prevention means that attention must be concentrated at the beginning of a process i.e. at the input end, not after output.

12. The control of quality can be carried out only at the point of operation.

13. Quality control is the employment of techniques to maintain the quality of a product/process/service. It helps find and eliminate causes of problems.

14. Quality assurance is the activity of making sure that the system works. It is a systematic way of preventing quality problems.

The last two chapters have emphasized that achieving quality in any organization depends on the relationship between suppliers and customers. The vital function which must be worked out between them is the **process** by which the **inputs** can be transformed into **outputs** that are

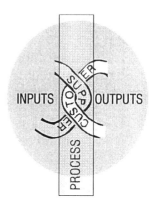

ORGANIZATION

EXTERNAL SUPPLIER

EXTERNAL CUSTOMER

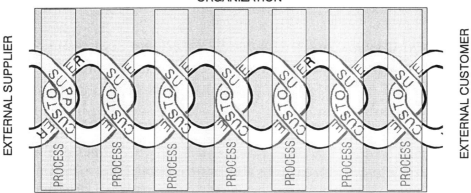

The suppliers and their customers, internal and external, form links in supply chains which produce the organization's final outputs for the outside customers. These chains are the core of a **model** which can be used for all organizations, regardless of their size or type.

OUTSIDE SUPPLIERS

OUTSIDE CUSTOMERS

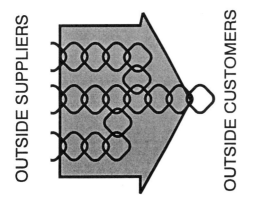

Although any given product or service will have several related chains contributing to it (as on the left) and, although there are typically a number of customers with varying requirements at any given time, for the purposes of developing a simple model let's suppose that there is one supplier/customer chain to represent them all:

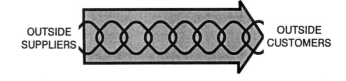

OUTSIDE SUPPLIERS

OUTSIDE CUSTOMERS

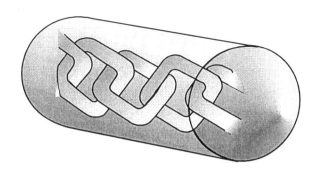

*Here it is in three dimensions, an active force
inside the organization, its objective - to deliver
the product or service to the outside customer.
However, it will not be able to do this if it does
not have an efficient means of propelling this
force, or of continuing to do so.*

*The supplier/customer chain is vital, but, on its
own, it will not be enough. It will need a support
mechanism..*

It will need...

Systematic planning

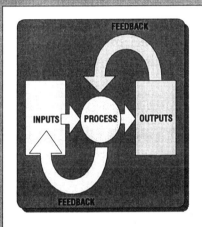

**Tools for measuring,
delivering and
sustaining quality**

**Organizing
for quality &
developing teams**

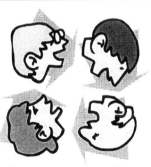

**Communication
between all
parts of the
organization**

**Commitment of
the organization to
a TQM approach**

**Recognition and
perhaps change of
the organization's
culture &
environment**

**All these supporting elements are essential ingredients
in making a model for Total Quality Management which
will be relevant to every business or organization, whatever
its size or function.**

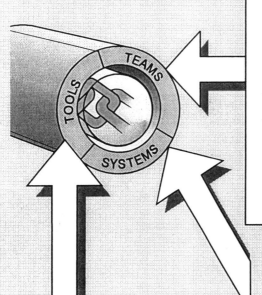

Teams

This element, explored in more detail in Chapter 8, deals with organizing for quality performance management, action-centred leadership and, in particular, the development of teams.

Systems

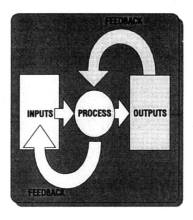

Systems are needed to ensure that each process will convert inputs into satisfactory outputs. Systematic planning is a basic requirement for effective quality management. The design and implementation of systems which guarantee quality performance are discussed in Chapter 6.

Tools

This element consists of the tools and techniques needed to deliver and sustain quality. Tools include the use of accurate measuring devices for identifying opportunities for improvement and for comparing performance against internal and external standards. This is covered in Chapter 7.

**These three elements affect the day-to-day
application of quality management.
They form the inner of two bands...**

22

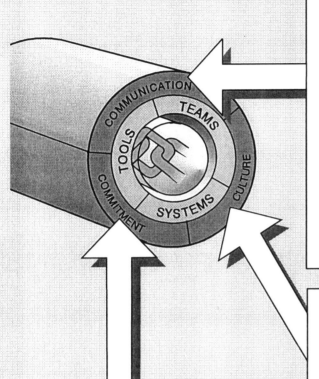

Communication

Internal communication between individuals and departments is an essential conduit for performance management and teamwork. It is dealt with in more detail in Chapter 5. It is also essential to establish good communication links with suppliers and outside customers.

Culture

The organization's ability to accept change, its morale and the internal working relationships are some of the factors which are crucial to the implementation of total quality management, as are its relationships with the wider business environment. This fundamental area is covered in the next chapter.

Commitment

Commitment to quality must come from top management, but any attempt to adopt a quality system will fail without the wholehearted support of the rest of the organization. We look at the importance of commitment in the next chapter.

The six elements making up the support mechanisms of the central core of the TQM model are the essential ingredients for a successful programme and form the content of the rest of the book.

Chapter 4 will begin to look at them in depth, beginning at the foundations, and asks "Are you ready for quality?"

1   The customer-supplier chains (internal and external) form the core of the TQM model, usable in all types of organizations.

2   The chains need a support mechanism comprising systems, tools and teams. These in turn are supported by communication, commitment and the culture within the organization.

3   Communication is dealt with in Chapter 5.

4   Systems are described in Chapter 6.

5   Tools are covered in Chapter 7.

6   Teams appear in Chapter 8.

7   Implementation of the TQM model is shown in Chapter 9.

8   Commitment and Culture are the subject of the next chapter.

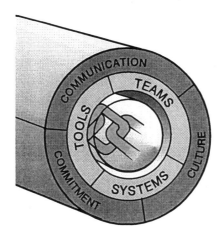

4 Commitment & culture

Ardwick and Hardcastle have been my life

The fundamental requirement for making total quality work is COMMITMENT.

But, what is commitment really?? Take this old gentleman.

Mind you, nothing's been the same since young Mr Hardcastle changed the horses for that contraption in '37

'Commitment' to the status quo is not commitment in the total quality sense . Neither is unquestioning loyalty alone...

MY COMPANY RIGHT OR WRONG

...because when the managing director wakes from his nightmare that the company is vulnerable to a hostile takeover...

...he knows that many changes will have to take place in order to survive. He will work long and hard to devise a plan that will save the company, but he knows it will stand no chance of success unless he can carry the work force with him.

But the problem is - how can he motivate his colleagues and employees to the point where they are committed to make his changes work?

He can appeal to his fellow directors relatively easily. But what about the others?

The managing director speaks...

I have a VISION!

...a vision of OUR COMPANY, successful beyond our WILDEST DREAMS!

Bringing to the buying public the BEST POSSIBLE PRODUCT at the BEST POSSIBLE PRICE!....

...a company where EVERYBODY is HAPPY AT WORK!

But things will have to CHANGE!

Go forth and SPREAD MY WORD!

A VISION!

WE HAVE A VISION!

The personnel director becomes involved.

Tell the lads

You see, we have this vision...

Er...

...so remember, chaps, VISION's the word

What was all that about?

Dunno, something about double vision or something

I think they got the message...

This is emphatically NOT the way to implement total quality.

Let's start again...

The MD tried to implement the whole of his TQM plan from the **top down**

MD ↓

Work Force

The personnel director was given the impossible job of managing a

CHANGE OF ATTITUDE

The MD's dream is unlikely to be shared by the workforce..

Even when supported by new programmes for change, purely top-down 'quality management' will not work.

We have a VISION

We do not have a VISION

and the most probable outcome of this line of action is confrontation and strife. The workforce are likely to be suspicious of change implemented from the top.

There will tend to be a lack of understanding of what needs to be changed, and how those changes are to be brought about.

It relies on the belief that organizational changes, mission statements, programmes and courses will themselves make transformations.

The more the change is imposed from the top the less chance that the change will be successful

Wear 'em or else...

So, make me

- particularly if just one group in the organization 'owns' the programme.

It is far more likely that people will take responsibility for change in areas of their own competence, particularly areas over which they have control.

So the problem becomes one of motivating changes to their **processes** and aligning employees' roles and responsibilities accordingly.

An organization will change through changes in its processes

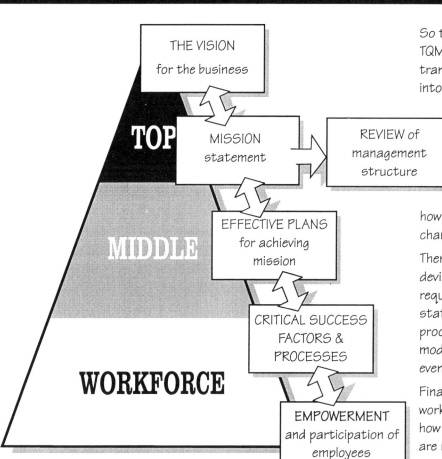

THE VISION
for the business

TOP

MISSION
statement

REVIEW of
management
structure

MIDDLE

EFFECTIVE PLANS
for achieving
mission

WORKFORCE

CRITICAL SUCCESS
FACTORS &
PROCESSES

EMPOWERMENT
and participation of
employees

So the successful integration of TQM into a business depends on translating the corporate vision into practice. This is done in a series of steps involving, first, top management whose job will be to embody the vision into a mission statement (see page 30) and to consider how the organization will be changed.

Then middle management must devise ways of meeting the requirements of the mission statement by identifying key processes which may need modification and improvement, or even redesign.

Finally, the expertise of the workforce will be needed to work out how the processes for which they are responsible can be improved to meet the new demands.

Everyone in the organization must be motivated to bring about change in his or her area of responsibility

The principles of the management of change are universally applicable

Once upon a time there was a certain head teacher in a certain school

This is a terrible school

In times gone by it had had a high reputation, but now the children were unruly and the staff apathetic.

Even the old hands had given up hope.

I've been teaching here twenty years and it's gone downhill since its days of excellence. These days nothing seems worth the effort

I'm sure there's nothing wrong with the kids...

I have a vision of a far better school, but I can't get the staff to share it

The teachers and the pupils have a low expectation of success at school. They are reconciled to the idea of failure at learning and teaching, and the head teacher seems unable to get the staff to change their attitude, despite his pleas at their regular weekly staff meeting...

We can get more out of these kids

When's the last time Kevin taught a class?

Not my hooligans

All right, you lot. Two plus two

We haven't done that

Don't need two plus two to be an aero engineer

Oh no, not the dreaded two plus two again!

There seemed no escape from the spiral of failure which breeds failure, when, suddenly the Angel of Quality appeared...

Translate your vision statement into processes

Eh?

The head teacher learned of the need to translate his vision into a mission statement acceptable to his staff, and then identify the critical success factors and key processes which would produce results.

Aim for the key processes that will affect the school's corporate identity and performance

Hang on, I'm no typist

School Uniform

...and identify the key elements that need to be changed. Come on, we're going to look at another school

The guru set up an appointment with another school in a similar environment, but with a much better track record...

This school used to be bottom of the ladder, but Phil Banks has transformed the place. This is where I leave you - good luck!

TOUGH STREET COMPREHENSIVE SCHOOL

Hi, Kevin, I'm Phil. I hear you're having our old problems. Let me show you around

This is the Project Room. The staff work in teams to set the senior kids projects covering a range of subjects, and the kids work them out unsupervised. Now let's look at the Computer Room

These first-years are learning a programme called C++. Mind you, we've got a cracking maths teacher

The Tough Street head teacher explained to Kevin how his policy of constant change and improvement kept the enthusiasm of the staff and pupils constantly alive.

No kid wants to end up a failure so, once they're on your side, it's a kind of spiral of improvement

Sure, but first I've got to get the staff to go along with me

I must have a meeting with my department heads right away

Here are some guidelines

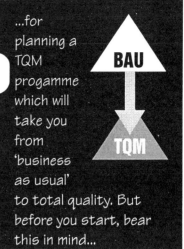

...for planning a TQM progamme which will take you from 'business as usual' to total quality. But before you start, bear this in mind...

When you have finally worked out your programme for total quality, it has to be introduced gradually.

A slow, planned, purposeful approach causes gradual change to take place so that 'business as usual' **becomes** total quality management.

To engage in TQM all at once, from top to bottom, will result in disaster, with people not knowing what to do and processes breaking down throughout the organization.

Now lets look at how you plan for TQM.

The first step is to gain commitment to change from the top team

The staff meeting begins...

For the sake of the children and our own self respect, something must be done...

Absolutely. I feel I can do better, and so can my department

I agree. We have to make a plan we can all subscribe to

Yes, it's about time we stopped feeling sorry for ourselves

About time. When can we start?

Um...

Next, develop a shared 'mission' and establish what change is required

Then define the distinctive competences of the organization...

...and identify areas of weakness

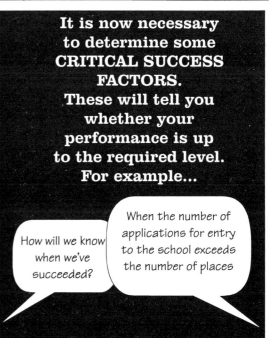

It is now necessary to determine some **CRITICAL SUCCESS FACTORS.** These will tell you whether your performance is up to the required level. For example...

The next step is to define MEASURABLE OBJECTIVES which are key indicators of performance

We have to improve our exam results - and go on improving them

and breaking down critical success factors into key processes

This means staff going on in-service courses, learning from others, throwing away tired old teaching methods, feeding back results, forming teams to combat weak performances...

Finally, break down the critical processes into tasks

I'd like your ideas on paper by the end of next week please. Chris, you wanted to see me?

Teaching's a mug's game, Kevin

Thought of early retirement, Chris? The LEA has a scheme...

Provoking change can also have the effect of removing dead wood

And so a new mathematics teacher is appointed

What ideas do you have about teaching maths to 13-year-olds?

Wow, LOADS! You just have to relate to their experience and imagination...

So, how does the balloon's air speed differ from the aeroplane's?

Please, miss! Balloon's air speed is always zero!

$(m)x-y$

One year later...

The school's performance has improved at increasing speed as staff regained their self-respect and found the scope to use

I am really sorry to disappoint you, Mrs White, but our intake quota has been full for months

OXFORD CANDIDATES

their imagination, while the pupils become too absorbed in the work to think of misbehaving. Academic results shot up and the school is now much sought after by aspiring parents.

But for the seeds of change to grow, the ground must be fertile...

However great the commitment of the chief executive or his staff, if the culture of the organization is not one where ideas are allowed to appear and develop, total quality management will fail...

The production manager speaks...

Yes, Mr Bligh

I run a tight ship, Jenkins

I won't have slackers on this site. That man over there is STANDING STILL!

Yes, Mr Bligh

If he doesn't have a good excuse, fire him

Meanwhile

This packing roll is slow as sludge. I've got an idea which would make this three times faster

Watch your lip, Ernie, you know the old man don't like being told his job

Mr Bligh..phone!

Pah!

Yes, Mr Bligh

It is a customer

Listen, Bligh, the last delivery from your so-called production company was late, and as you are the so-called production manager, the can is yours, right?

Yes, Mrs Morgan...No, Mrs Morgan...No, I assure you...No, I will see to it...No, you won't have to cancel...No, please accept my most sincere...Goodbye, Mrs Morgan..

Jenkins, get me whoever runs the packing department!!

Yes, Mr Bligh

I've been tellin' you lot for ages that wrapper is a load of rubbish but you won't heed

Well it's no concern of yours now because you're OUT!

That's right, Mr Bligh

SUPPORT OUR ERNIE

Yeah!

Yeah!

ERNIE OUT – EVERYBODY OUT!

Yeah!

As Ernie's shop steward I will see that he gets full union representation about this, and having regard for the full facts of the case I have no alternative but to recommend to the union that it withdraws its labour forthwith

Organizations have constantly to plan for change to meet changing demand; but without a sympathetic culture, no one will want to feel involved and no change will be possible – particularly from the people who will have to implement it - the workforce.

Meanwhile, that night, Bligh slept, untroubled by the chaos he had caused.

Suddenly...

Shame on you, Bligh!

"Your tyranny could close the company," said the ghostly guru...

"...you will be spurned by your colleagues and probably lose your job..."

I don't even smoke

"...and you will be reduced to picking up cigarette ends from the gutter!"

"Have you never heard of teamwork? To get the best from your workforce you have to work together - not against each other. You're all on the same side, remember?"

To get any improvement you must change the CULTURE of your organization so that people will feel a sense of COMMITMENT

Anything, sir, anything! But how do I start? And what do you mean by culture?

Let's go back in time, but we'll leave your yes-man at home

It's the wrapping shop

The culture of an organization is made up of the shared beliefs about how business is conducted, how employees are treated and how they behave

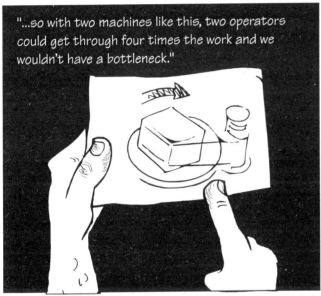

All members of the organization, from top management down, are responsible for improvements in the job they do, and for making sure their supplier/customer process has zero defects

Implementing TQM means creating new responsibilities in a process-driven environment

We have to change. Now what kind of company do we want to be?

Er... I give up

1- The mission statement

The first step is for the organization to state clearly its beliefs and objectives. The mission statement is the means of expressing them. It should:

- define the business of the organization & its role;
- state its commitment to effective leadership and quality;
- identify target sectors, relationships with customers;
- place the organization in market terms;
- state its distinctive competence, compared with similar organizations;
- give a statement of future plans under consideration;
- give a commitment to monitoring customers' needs & to continuous improvement.

2 - Strategies and plans

The organization must create strategies to gain its objectives and secure its position in the market place, and then it must develop plans to realize its strategies.

Commitment to TQM will be higher if employees could help senior management in the planning process.

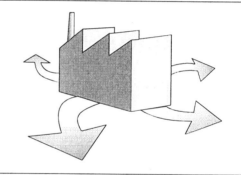

3 - Critical success factors and key processes

Next, it will be necessary to identify the critical success factors - the most important sub-goals of the organization. CSFs are the details of what must be accomplished to achieve the mission.

The CSFs are followed by the key processes - the activities that need to be done particularly well to ensure success.

4 - Reviewing the management structure

After defining the corporate objectives and strategies, CSFs and key processes it may be necessary to review the structure of the organization to make the new plans work.

Managerial responsibilities may have to be redefined and new procedures implemented. This review should also include an organization-wide establishment of a process quality improvement team structure.

5 - Empowerment and participation

Management of the empowerment of employees subdivides into managing communication, attitudes, abilities and participation.

Managing communication:

1 communicate with employees at all levels of the organization;

2 take appropriate action after communication;

3 encourage good communication between all suppliers and customers in the chain.

Managing attitudes:

promoting self-awareness. The key attitude for all members of the organization must be:

I must know who my customers are, what they need from me; I must be aware of how well I am satisfying their requirements and take whatever action is necessary to satisfy them fully. From this attitude comes its corollory: I must, as a customer, make my requirements plain to my suppliers and provide feedback on their performance.

Managing abilities:

Every employee must do the job needed of them - but first it's necessary to decide **what's** needed. Changes in processes mean changes in what's required, which will mean learning new skills, or acquiring new knowledge. So **training** must be **built-in** to the employee's job description; and it must be planned and continuous.

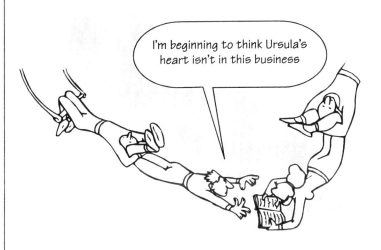

Managing participation:

All employees must be trained in the basics of disciplined management, i.e. they will do what they say they will do. They must be trained to **evaluate** any situation and define the necessary objectives, **plan** to achieve those objectives, **do** (implement their plans), **check** that the objectives are being achieved and , finally, **amend** or take corrective action if they are not.

The spiral of never-ending improvement

The sequence -
EVALUATE, PLAN, DO,
CHECK and AMEND - is a
continuous process and
will lead to a climate of
continuous improvement
within the organization.

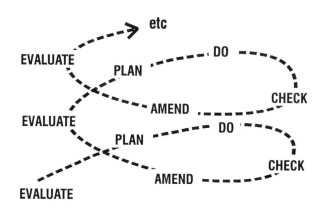

Ten points for senior management

The TQM model shows how all the factors affecting an organization's performance relate to each other. Here, to finish the chapter, are ten points - a summary of the various messages from gurus which can be used by senior managers when establishing a policy based on quality.

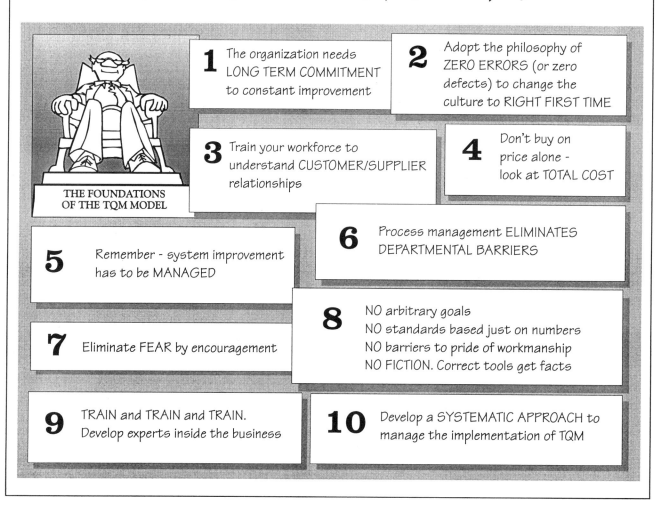

THE FOUNDATIONS
OF THE TQM MODEL

1 The organization needs LONG TERM COMMITMENT to constant improvement

2 Adopt the philosophy of ZERO ERRORS (or zero defects) to change the culture to RIGHT FIRST TIME

3 Train your workforce to understand CUSTOMER/SUPPLIER relationships

4 Don't buy on price alone - look at TOTAL COST

5 Remember - system improvement has to be MANAGED

6 Process management ELIMINATES DEPARTMENTAL BARRIERS

7 Eliminate FEAR by encouragement

8 NO arbitrary goals
NO standards based just on numbers
NO barriers to pride of workmanship
NO FICTION. Correct tools get facts

9 TRAIN and TRAIN and TRAIN. Develop experts inside the business

10 Develop a SYSTEMATIC APPROACH to manage the implementation of TQM

1. The fundamental requirement for making total quality work is senior management COMMITMENT.

2. Real commitment generates a vision which is meaningful and achievable.

3. Any changes should not be implemented top down alone but must involve everyone.

4. The vision and mission for the organization must be translated into effective plans for implementation. This means identifying the critical success factors (CSF's) and key processes.

5. These principles of managing change apply to all types of organization.

6. The culture of the organization must support the senior executives' commitment to TQM. A climate of fear will not do so.

7. Involvement of the work force is essential to achieving lasting improvement or change.

8. The five steps to integrating TQM into the business include consideration of the mission statement, strategies and plans, critical success factors and key processes, reviewing the management structure, and empowerment and participation.

9. Ten points to guide senior management refer to: long term commitment; zero errors; customer/supplier chains; total cost of purchase; system improvement must be managed; elimination of departmental and pride barriers, fear, arbitrary goals, standards based only on numbers and fiction; developing experts through training; and a generally systematic approach.

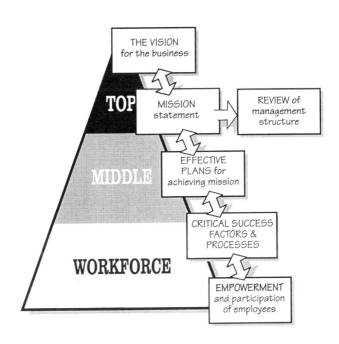

5 Communication

Bright ideas by senior executives do not communicate themselves

The essence of changing people's attitude to quality is <u>gaining acceptance</u> for the need for change

A fundamental tenet of communication

> It sounds like plain common sense, but it's surprising how bad communication is in some organizations.
>
> Good communication is the KEY to quality improvement

Communications strategy should be built into the organization, from the top down

```
TOP MANAGEMENT ── QUALITY COUNCIL ──┬── TQM CO-ORDINATOR
                                    ├── MIDDLE MANAGEMENT
                                    ├── SUPERVISORS
                                    └── OTHER EMPLOYEES
```

A communication structure could be organized as follows: top management appoints a quality council consisting of employees from all levels of management and supervision. The council is responsible for co-ordinating communications policy throughout the organization, but the task of keeping the communication channels open is in the hands of a TQM co-ordinator.

The first communications step in a quality improvement programme is a statement from top management

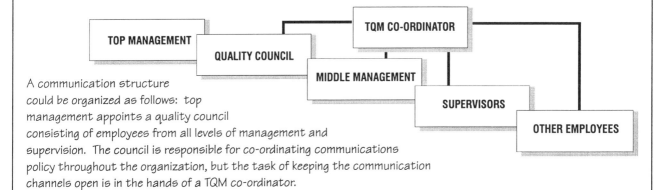

For example:

The board of directors believe that the successful implementation of Total Quality Management is critical to achieving and maintaining our business goals of leadership in quality, delivery and price competitiveness.

We wish to convey to everyone our enthusiasm and personal commitment to the total quality approach, and explain why your support is so important in our mission for process improvement. We hope that you will become as convinced as we are that process improvement is vital for our survival and continued success.

We can become a total quality organization only with your commitment and dedication to improving the processes in which you work. We will help you by putting in place a programme of education and teamwork development, based on process improvement, to ensure that we move forward together to achieve our business goals.

> That's the chairman, Mr Dobson!

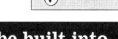

> Aye! The Old Man's right enough

> Some of you joined the firm when I did as a boy fifty years ago. I'm telling you lads and lasses we need to work together to stay where we are, on top.
> So we'll do our bit if you do yours

43

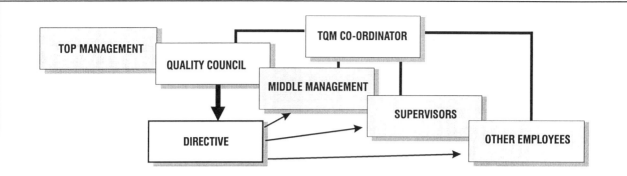

The TQM co-ordinator should then assist the quality council to prepare a directive, to be distributed to everyone in the organization. The directive should be signed by every leader of a business unit, division or process and include the following:

Contents of the directive

- **The need for improvement**
- **The concept of total quality**
- **The importance of understanding business processes**
- **The approach that will be taken**
- **Individual and process group responsibilities**
- **The principles of process measurement**

The directive should be disseminated by all the conventional communication methods at the disposal of the organization

These are discussed on pages 47 to 49.   In any organization there is always opposition to change, so:

First-line supervision will need to review the directive with all the staff.

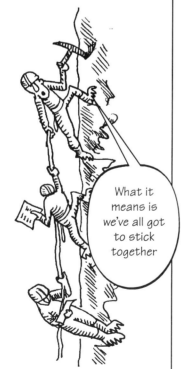

What it means is we've all got to stick together

Everyone must know exactly what's going on, otherwise any new initative will be met with suspicion and resentment.

What are the bosses up to now?

It is vital to get everyone to understand the aims and benefits of TQM, otherwise any attempts to facilitate change will meet with opposition.

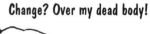

Change? Over my dead body!

...and there must be regular meetings between employees and their managers or supervisors to communicate general information about TQM.

Different groups will have different initial attitudes towards TQM

FOUR AUDIENCE "GROUPS" - AND THE ATTITUDES THEY BRING WITH THEM

What's good for us is good for the company

TOP MANAGEMENT

TQM is an OPPORTUNITY for the organization and for US

You know who'll have to do all the work, don't you?

MIDDLE MANAGEMENT

TQM is YET ANOTHER BURDEN

They've got nothing better to do on the 5th floor

SUPERVISORS

TQM is ANOTHER FLAVOUR OF THE MONTH

Just ignore it and it will go away

OTHER EMPLOYEES

TQM is TOTALLY IRRELEVANT

The job of communication is to ensure that each group sees TQM as being beneficial to them

Of course, organizations come in all shapes and sizes

so it will be up to each one to develop a communication system best suited to itself. A crucial aim of the communication process is to encourage all employees to stop thinking purely in departmental terms and start thinking in terms of the whole organization.

This will mean breaking down departmental barriers by concentrating on **processes** rather than 'departmental' issues.

COMPETITION

The real conflict should lie **outside** the organization

The communications strategy for the organization will have succeeded when everyone in it is convinced of the benefits TQM can bring to them as well as to the organization.

TQM means for

MIDDLE MANAGEMENT

I can manage more successfully - therefore I can earn promotion

TQM means for

SUPERVISORS

a more motivated workforce - therefore my job is made easier

TQM means for

OTHER EMPLOYEES

more job security and better training prospects for me

There are four principal ways of communicating

Verbal

Verbal communication, either between individuals or groups, using direct or indirect methods, such as public address systems, broadcasting or tape recordings

Advantages:
- Direct impact and feedback
- Permits plain language
- Permits presenter to check assimilation
- Allows presenter to gain audience commitment

Disadvantages:
- Depends on presenter's communications skills
- Uses only one of audience's senses
- Each presentation requires time to prepare
- Uniformity of content/understanding uncertain
- Time consuming & most effective for small groups

Written

Written communication in the form of notices, bulletins, information sheets, reports and recommendations

Advantages:
- Same message to all. Speed
- Everyone can receive message at the same time
- Can deal with large audience quickly
- Simultaneous circulation by various means
- Backup to complicated verbal communication
- Information exists in recorded form

Disadvantages:
- No guarantee of receipt or understanding
- Ambiguity of written language without feedback
- Impersonal, inanimate, reduces participation

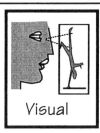

Visual

Visual communication such as posters, films, video tapes, demonstrations, displays and other promotional features. Some of these also call for verbal & other audio communication

Advantages:
- 75% of all information received is visual
- Possible to convey movement
- Graphical representation possible
- Participation possible (e.g. via flip chart)
- Vehicle for non-verbal information

Disadvantages (when indirect):
- No guarantee of receipt or understanding
- Impersonal, reduces participation

By example

Example, through the way people conduct themselves and adhere to established working codes and procedures, through their effectiveness as communicators

Advantages:
- Role model provides target for performance
- Standards in attitude reinforced
- Rules & procedures exactly demonstrated
- Housekeeping & hygiene standards kept
- Feedback strong

Disadvantages:
- Usually applicable to small groups
- Communication process is protracted

Good communication involves using the most appropriate medium...

...and plain language from immediate superiors

Some methods of communication for quality improvement

Each organization develops a system of internal communication best suited to its needs. In some, the main channel is the telephone, in others, written memos. Here is a by-no-means exhaustive list of communication methods in common use.

SUGGESTION SCHEMES

Where such schemes already exist for general matters affecting the organization, a period may be set aside for suggestions which deal specifically with quality topics.

These should be used sparingly to gain maximum impact. Presentations should be given for the best suggestions, with the appropriate publicity.

DEPARTMENTAL TALK-INS

Known sometimes as 'huddles' or team briefings in the United States, this method gathers together people from the same department for brief periods to discuss quality issues relevant to that group.

Time is usually short - sometimes a tea or coffee break, or at a shift change, so an agenda should be prepared in advance and the sessions kept short by rapid coverage of the points.

POSTER CAMPAIGNS

Posters can be a very effective part of the quality communication message if used from the very beginning.

The first posters should be simple and straightforward, with direct messages like: "Quality Starts Here", "The next person who checks your work will be your customer" and "Get it right first time and avoid waste".

Cartoons and drawings can often add impact.

It will be useless to stick posters up at random. A poster campaign must be carefully planned, organized and managed. Ask yourself "What do I want to achieve with the campaign?" and choose the most relevant posters available.

Choose poster locations with care. They should be well-lit and not interfere with movement, but should be in a prominent position.

They should be placed at eye level and not cluttered with other posters or wall publications.

'Home-made' posters are often better than impersonal commercial ones, particularly if there has been a competition within the organization for the best quality poster.

The message in the posters should change as employees' awareness of the campaign develops. Refresh the communication messages regularly - say every three months.

INDUCTION AND VOCATIONAL TRAINING

'Quality consciousness' begins when a new employee enters the organization. Induction training alerts people to the organization's requirements, code of practice, conduct and the quality culture. Their experience of quality during the induction process should make them want to learn more of it.

Vocational training for specific jobs should satisfy the employee's interest in quality gained during the induction period. Quality training should be integrated with vocational training by relating a quality approach with the consequences of its absence.

POINT OF WORK REMINDERS

During one energy crisis the use of 'Save It' stickers throughout organizations caused all employees to be constantly aware of the need for conserving energy.

This idea can be used for highlighting special problems and for encouraging careful working practice, especially where these have been neglected in the past.

COMPETITIONS

A competition may be at the organizational level with, say, a determined effort to win a national or continental 'quality award' such as the Malcolm Baldrige National Quality Award in the USA, or the European Quality Award (see Chapter 6); or it could be an internal competition, either company-wide or on a departmental basis.

Quality competitions are no substitute for training: their purpose is simply to raise the level of awareness of the need for quality. If they fail to generate interest and improvement they are useless.

Many quality competitions are based on error or defect rates over a certain period. To make the competition even when it involves departments with dissimilar risks, some modifying calculation must be brought in to make the competition fairer. For example, results could be based on the percentage reduction on each department's defect rates compared to a previous period; or error frequencies with a 'handicap' based on the relativity of the risk weightings between the two departments; or each department is assessed by a team on a set number of occasions at random intervals. Marks are awarded according to the level of customer satisfaction recorded.

PRIZES, FORMAL PRESENTATIONS and NEWSLETTERS

In organizations where presentations to process quality teams, quality improvement teams or quality circles are part of the recognition process, it is a good idea to award some sort of certificate of recognition for people to display in their workplace.

Prizes can vary from copper-etched plaques to weekends in Paris. If lunches and presentations from senior executives of other organizations are used to deliver the prizes, the commitment and support from the top are underlined.

Photographs and reports of such award ceremonies play a part, keeping the reward process going through company newsletters - the items should be particularly eye-catching - and might even merit a page of their own, to be regularly included.

Fred is marvellous!
That's his eighth teddy bear
this year!

DEMONSTRATIONS AND EXHIBITIONS

Static exhibitions of certain aspects of quality can be a focal point of interest and a powerful way of making an impact.

OPINION OR ATTITUDE SURVEYS

In some companies, employee opinion or attitude surveys are conducted by questionnaire as part of TQM. If these are designed carefully and distributed efficiently they can measure the employee's perception of the programme.

One danger with this way of gathering information is the development of complacency if the results are consistently positive. TQM demands continuous improvement, and each achievement should set targets for further improvement in the future. Another danger is that TQM raises expectations, and the second survey - in a series begun before TQM - may indicate that things have worsened. What has actually happened is that the awareness of problems has been raised through the education and training process, and the questionnaire responses are more critical.

In no organization are communications perfect.
In most organizations communications need a lot of improvement.

5 Summary

1   The essence of changing people's attitude to quality is gaining acceptance for the need for change, through good communications.

2   The communications strategy for quality should be built into the organization from the top down, through a quality council and the appointment of a TQM co-ordinator.

3   The first steps in a quality improvement programme are a statement from top management and a directive, which are communicated to everyone so that they may understand the benefits for them of TQM.

4   There are four principal ways of communicating: verbal, written, visual and by example. Each has its own requirements, strengths and weaknesses. Good communication involves using the most appropriate medium.

5   Ten methods of communication for quality improvement are:

suggestions schemes;
departmental talk-ins;
poster campaigns;
induction and vocation training;
point-of-work reminders; competitions;
prizes/formal presentations; newsletters;
demonstrations/exhibitions; and
opinion/attitude surveys.

6   There are three basics of good communication:

keep it simple;
make as much of it as possible face-to-face;
use first line supervision and plain language to communicate.

7   In most organizations, communications need a lot of improvement.

At this point I must confess to two failings. The first is a shockingly bad memory for dates...

Naturally, I keep a diary, and so does my secretary

...so as long as he remembers to check his diary with mine we're OK!

Manager's entries **Secretary's entries**

Diary 1	Diary 2

Daily consolidation at 9 am

Diary 1	Diary 2

This elementary activity, familiar to countless managers, is a SYSTEM. It is a process which minimizes the error factor in recording appointments.

My other failing is not so easily resolved.

I have yet to produce a satisfactory soufflé.

However hard I try, my soufflés are very variable!

The only consistency I could achieve was that each soufflé could be relied upon to be different from the previous one.

I decided to give up, and buy soufflés instead of trying to make them.

The chef-proprietor was Alain, who waited upon us personally.

Monsieur, Madame: I welcome you to Le Roi des Soufflés!

KITCHEN

And so, my wife and I are to be found one evening enjoying an apéritif before entering a restaurant specializing in soufflés. I remember thinking that the job of soufflé chef must qualifiy for danger money, since the chances of success every time must be small.

Un...deux...trois...

Soufflés are fiendishly difficult to get right, requiring perfect preparation and split-second timing...

...voilà!

How could the chef make perfect soufflés **time after time** for a restaurant full of customers?

For me, la cuisine has been never a problem - from the beginning it was the play of the child, so to speak. The soufflé has been for me the specialité, and so I from an early age regarded my destiny as that of the chef-soufflé

Merveilleux!

Le ROI des SOUFFLÉS

Prochain

Regrettably, success brought its travails, and I was unable to satisfy the enthusiasm of my public for les soufflés. Each soufflé takes almost an hour to prepare, so it became necessary for me to engage collaborators. But, malheureusement, none had the genius of Alain for the preparation of the soufflé and so I was obliged to write down the method to be followed without question. Only in this way could I be sure that the soufflé made in my name should carry the requisite quality.

In this country I achieved even greater success, even though your chefs are not, regrettably, French and some of the ingredients, dommage, are Italian. Carefully prepared procedures, in which I involved the chefs, have overcome these grave deficiencies, with the consequence that I am released from the necessity of preparation myself which allows me to participate in our agreeable tête-a-tête.

I can give you a glimpse only of our documentation. Regrettably, to reveal more is not possible.

The Souffle King
Souffle Fromage
Oven at constant 190C

Ingredients	Preparation	Timing	Team
Eggs 6	Separate eggs	00.00	Team 1 In advance Dish 1 whites Beat until stiff Dish 2 yolks
Milk 25cl Flour 20g Parm. 20g Salt 8g Pepper 8g	Mix roux add cheese	05.00	Team 2 Dish3 100C Add yolks from dish 2

Tiens! C'est FROID!

There have been changes and improvements, of course, over the years. Gaston, a talented sous-chef in one of my restaurants, discovered that a refrigerated spoon, when preparing the egg-whites gives greater endurance to the rigidity of the soufflé, so we have amended our process to accommodate this. And so we have a system which is at once proof from the fools but flexible to take account of ameliorations which can be made as we proceed.

Then the improved process becomes once more the one we follow.

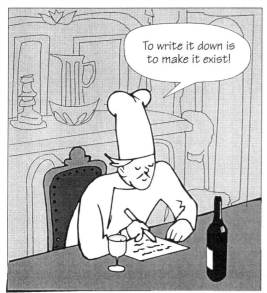

To write it down is to make it exist!

Write down what you do
↓
Justify it
↓
Do what is written
↓
Record what you did
↓
Review it
↓
Revise what you will do

Improvement loop

And so my advice to the makers of soufflés, like yourself, or to anyone else for that matter, is to write down your procedure very carefully, and follow it. Allow improvements, indubitably, but then the documentation must be changed.

So, here we have a system with an **improvement loop** built in. Because of its capacity for change, this guarantees a **continuous system for improving quality.**

Many organizations have established systems which work perfectly well, and the temptation is to rely on informal ways of effecting improvement - but sustained and controllable improvement will not be possible unless the system and its changes are documented, practised, audited and reviewed.

Adrian never changes his act without telling me first!

But even the best documented quality systems will not work if they are ignored. This will happen if the written system does not evolve from what actually happens or if the people carrying out the procedures are not properly consulted.

The door kept banging so I found a use for our systems manual at last

Writing down your systems is no substitute for using them

In documenting a system you watch the process...

...and audit every stage, noting the items which go into the process - timing, division of tasks, technical specifications etc., and you end up with a written description of how the whole thing is accomplished.

55

The system is audited in practice and improved where appropriate - and the changes are written down as they are agreed.

The task of documenting a system should involve the people who have a responsibility for any part of it

In the operation of any process, a useful guide, and one which underlines the need for processes to be controlled by systems, is this:

No process without DATA COLLECTION

No data collection without ANALYSIS

No analysis without DECISIONS

No decisions without ACTION*

*which can include doing nothing

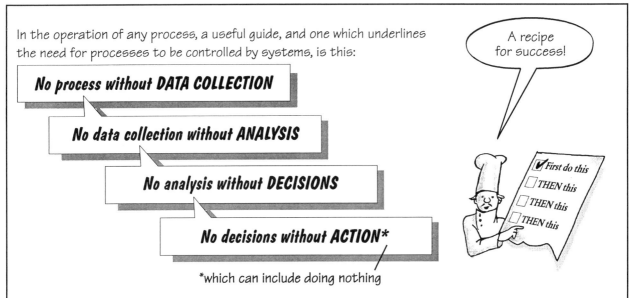

So a SYSTEM essentially consists of a series of check lists which should help the successful production of products or services with consistent quality and reliability. An agreement on system standards in all kinds of businesses has been formulated by the International Standards Organisation, under the generic number of ISO9000. System specifications in this set of standards have numbers which follow on - ISO9001, 9002 and 9003.

For a company's quality system to conform absolutely to customer's requirements it would have to endure frequent second party inspection. Third party certification to the ISO9000 series means that the company is assessed at regular intervals by independent experts instead of a long stream of customers. The company becomes a "registered form of assessed capability".

Organizations which operate according to accepted standards enjoy the trust of their customers and potential customers

Some misconceptions exist about the constraints imposed by a system like the ISO 9000 series - for example some people think that there are limits imposed on their operation, while another generally held view is that, because the firm applying for registration can specify their own standards, the system will condone rubbish.

Neither is true. There are no constraints on performance, and customer requirements must be met for the organization to stay registered. While the ISO 9000 series of standards help the quality of goods and services from suppliers and to customers to remain consistent, it does not monitor directly human relationships and teamwork. For a really comprehensive framework it's necessary to turn to a business excellence model such as the US Baldrige or the European Quality Awards. The European Award principle claims that customer satisfaction, employee satisfaction and impact on society leading to good business results are achieved through leadership-driven policy and strategy and good people, resources and process management. The links are shown in the diagram below, together with the weightings used in the award process. These emphasize the people-driven philosophy of the system.

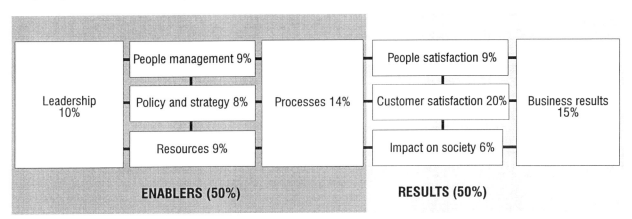

Many managers feel the need for a rational basis on which to measure TQM - they look for answers to questions like "Where are we now?", "Where do we need to be?" and "How do we get there?" The answers must come from **employees'** views, **customers'** views and the views of **suppliers**. Internal or **self-assessments** will provide an organization with vital information in monitoring its progress towards its goals and total quality. The external assessments used in the process of making the European Award are based on those self-assessments that are performed as pre-requisites for improvement.

1   A documented quality system which is followed by everyone helps to ensure customer satisfaction.

2   Improvements in processes should lead to changes in procedures and the documentation.

3   The quality system guidelines are:

> write down what you do;
> justify that what you do meets a good standard:
> do what is written;
> record what you did;
> review what you did;
> revise what you will do; and
> carry out the necessary corrective action for continuous improvement.

4   The documented system must evolve from what actually happens through the involvement of the people who operate the process.

5   A useful guide to the operation of any process is:

> no process without data collection;
> no data collection without analysis;
> no analysis without decisions;
> no decisions without action (which can include doing nothing).

6   The ISO9000 series of quality system standards is the basis of independent third-party certification.

7   The use of ISO9000 should not constrain performance, nor does it condone inferior products - the customer requirements should be demonstrably met.

8   More comprehensive frameworks than ISO9000 for "business excellence" may be found in the US Baldrige and European Quality Award models: customer satisfaction (20%), employee satisfaction (9%) and impact of society (6%) leading to good business results (15%) are achieved through leadership (10%)- driven policy and strategy(8%); and good people (9%), resources (9%) and process (14%) management.

9   Self-assessment against one of the quality award models can provide measured answers to questions such as: "Where are we now?", "Where do we need to be?" and "How do we get there?"

For outputs to meet the requirements of the customer, inputs must be defined, monitored and controlled.

I often feel that all organizations can learn some useful lessons from the world of sport

Fortunately, no one cares about the quality of my products or services

Not even me

I do not have the purely recreational side of sport in mind - quite the opposite.

I refer to the rigour and accuracy of professional sport. Take the 100 metre sprinter we see training here.

Ready... get set...

Bang!

click

click click click

His times are recorded with meticulous accuracy by his trainer.

They are using a measurement - in this case the times recorded by a stopwatch - to help improve the sprinter's performance. We'll see how recording data like this can come to the aid of other organizations in tracking down inefficiency and increasing profitability, like the case of the teabag factory on the right.

But first let's review a number of information-gathering tools at our disposal and see what function they perform.

An essential activity in improving processes is the collection of information

Here are some basic tools for collectiing and interpreting data - and the questions they answer.

1. WHAT IS DONE?

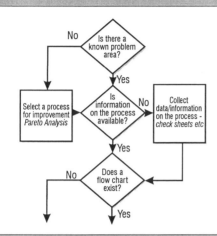

PROCESS FLOWCHARTING

The use of this technique makes sure that the inputs and flow of the processes making up the business are understood.

Process flowcharts show how complex even the simplest operation can be. Their use allows individuals to have a clear and shared idea of the process. In all but the simplest task, no one person is able to complete a process flowchart without help from others. Flowcharting is thus a powerful team-forming exercise.

2. HOW OFTEN IS IT DONE?

CHECK SHEETS or TALLY CHARTS

A check sheet is a direct way of observing and recording data, and a logical point to start in most process control tasks. It is a good way of data, which may be used in making decisions, or be of direct use to the job in hand. Merely to record how many times people slip on a banana skin in a certain location has no immediate value, but recording the incidence of skids in relationship to different locations might be.

4 points to remember:

❋ Select & agree on event to be observed.

❋ Decide on time period - how often & for how long.

❋ Design a simple, legible form.

❋ Allow enough time to complete recording & present analysis.

3. HOW DOES THE INFORMATION VARY?

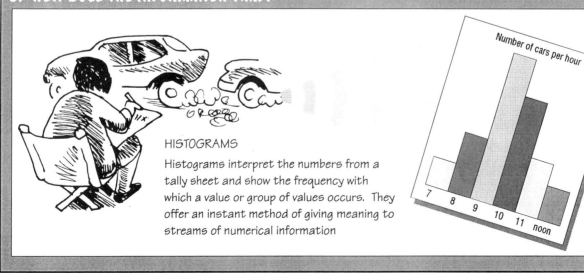

HISTOGRAMS

Histograms interpret the numbers from a tally sheet and show the frequency with which a value or group of values occurs. They offer an instant method of giving meaning to streams of numerical information

4. WHAT IS THE RELATIONSHIP BETWEEN FACTORS?

SCATTER PATTERNS

It's frequently useful to compare one set of data with another to establish a relationship between factors or parameters. For instance, there could be some connection between the height of a soufflé and the oven temperature over a certain period of time. Comparing data between the dimensions of the soufflé (y axis) and the temperature of the oven (x axis) might yield the pattern on the right and show the optimum temperature for success.

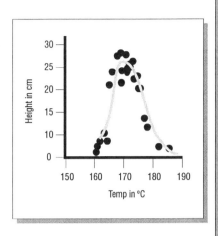

5. HOW IS THE DATA MADE UP?

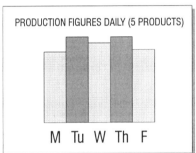

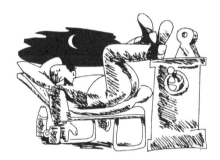

STRATIFICATION

Overall results shown as, say, a single chart or histogram can hide significant variations. Stratification, i.e. separating all the data into groups, can reveal weaknesses or aberrations. For example, the overall production figures of this firm look good, but when split into the three shifts, the story is somewhat different.

6. WHICH ARE THE BIG PROBLEMS?

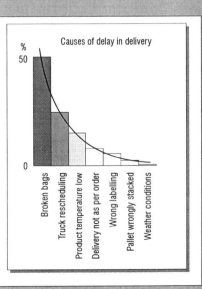

The assembly line's OK until it gets to this section

PARETO ANALYSIS

If the causes of some defect can be identified and recorded, it will be possible to tell what percentage defect and failure can be attributed to that particular cause. It will probably be found that about 80% of the errors, defects or breakdowns will stem from about 20% of the causes. The name comes from an Italian economist who found that 80-90% of Italian wealth was owned by 10% of the people. Pareto's Law applies to marketing too (80% of the business comes from 20% of the customers).

7. WHAT CAUSES THE PROBLEMS?

CAUSE & EFFECT ANALYSIS and BRAINSTORMING

One way of mapping the inputs that affect quality is the **cause & effect diagram**. Potential causes are shown as labelled lines entering the main cause arrowed line. Each line may have other lines entering it, as the principal factors or causes are analysed and divided into sub-causes and sub-sub-causes.

This process of analysis and division is called **brainstorming**. It's a technique for generating a large number of ideas quickly: each member of a group is asked to put forward all their ideas, however wild, about a given problem.

Every member has equal status, and the intention is to promote originality of thinking. All ideas are recorded and analysed later. Pareto analysis identifies the ideas most worth pursuing.

There are numerous variations on the basic brainstorming technique, to be found in *Total Quality Management* by John Oakland, published by Butterworth-Heinemann

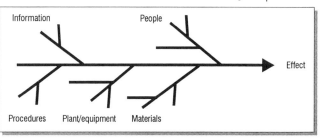

8. WHAT WILL HELP OR OBSTRUCT CHANGE?

FORCE FIELD ANALYSIS

Force field analysis is a technique used to identify the forces that either help or obstruct necessary changes and to plan how to overcome them.

A team describes the needed change and defines a possible solution

Pushing FOR	Pushing AGAINST
If we don't do it there will be a disaster	We are doing bits of it anyway
It will affect promotion	What about promotion?
Company committed to TQM	Existing management structure
Share ideas with union	Union may be against it
Improves efficiency	Current appraisal schemes
Quality action teams to be formed *with* equal opportunities	No equal opportunities

Having prepared the basic force field diagram (see left for example) it identifies by brainstorming the favourable/positive/driving forces and the unfavourable/negative/restraining forces.

These are placed in the diagram (FOR and AGAINST) then rated and evaluated for their potential influence on the implementation of change.

Then the team draw up an action plan to increase the driving forces and reduce the restraining forces.

9. WHICH ARE THE MOST IMPORTANT FACTORS?

THE EMPHASIS CURVE

This is a technique for ranking a number of factors in order of importance. The factors will not be able to be easily ranked in order of, say, cost or frequency of occurrence..

Each factor is compared in turn with all the others and the preferred factor is ringed. The factor gaining the highest number of rings becomes the most important factor in the item under consideration.

The technique consists of making a table with each factor compared against all the others in turn so that (in the case of the choice of features in a car) if there were 4 factors the table would look like the one on the right.

Item 1 Style	Compare with	① 1 1	2 ③④	
Item 2 Colour	Compare with	2 2	③④	
Item 3 Price	Compare with	3	④	
Item 4 Reliability		4		

Here, reliability (item 4) is ringed 3 times, so is the most important factor; price (3) is the next important - 2 rings, then style with 1. Colour comes last.

CONTROL CHARTS

These show when a process is going outside its specified tolerance or variation. They are based on the practice of taking samples at random intervals. The most frequently used control charts are **run charts**, where the data is plotted on a graph against time or a number of samples.

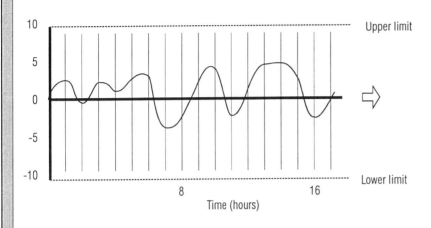

Variation of ingredient x in mixture

In one form of run chart, acceptable levels of variation (+ or - 10 units in this case) are set on a continuous chart plotted over time, and the process is monitored. No action is required until the plotted line goes either above or below the permitted variation as the mixture is within the agreed specification until that happens.

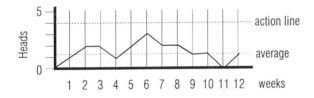

Heads on backwards - Ted assembly line

Another is a defect chart, showing the number of defects produced by a process. There is no 'acceptable' variation here, of course. The aim is zero defects over the entire period.

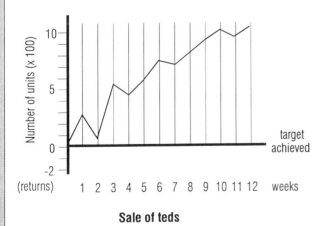

Sale of teds

Yet another is the CUSUM (cumulative sum) chart where a target value is subtracted from each successive sample observation and the result accumulated. This shows trends; so in this chart of sales figures, as the target is subtracted from each cumulative result the sales figures end up exceeding the forecast figures. If the sales figures were the same as the projected target figures for the period, the result would be a straight line - the target line (zero).

Trouble at the teabag factory...

Liz is the new management trainee in the production department. She is being shown round after work by her boss.

This is where we fill the teabags

Coo!

But we have a problem here. As your first job you might like to try to sort it out

5 tonnes of tea go into that hopper every week...

...but only 4.5 tonnes come out as bagged tea. We've tried all the obvious technical improvements. You have a go. I shall be watching to see how you get on.

NEXT DAY

You ladies operate the machine - YOU are the experts. Let's team up and try to solve the problem. Our objective is to reduce wastage by 50% in six months

If anybody knows that machine, it's us

OK, brainstorming - any ideas? Don't worry if they sound daft

Supplier is short-changing us

Teabags are leaking tea under the machine

Evaporation

There's no loss, just something funny with the measuring system

Well, I reckon we should be looking for spilt tea somewhere

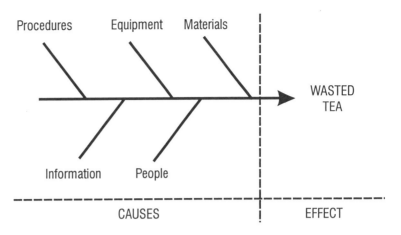

I agree. But bear in mind that the wasted tea is the effect, not the cause. We want to know why the wastage takes place so all your brainstorming suggestions can be grouped in this diagram according to the type of cause.

O.K., look at these piles of tea. This is what 2, 5, 10, 50 and 100 grammes look like...

...and for the next 2 weeks you are going to mark down each incident where tea is lost, whatever the cause,

College kids!

SO...

Should be 2 tonnes, Bert. I make it 1.92 - near enough

OK Len?

Oh no it's **not** near enough!

...and

SCRATCH!

BOFF

OOPS!

...and

...until, at the end of two weeks the operators had accumulated enough information on wasted tea to consolidate their results into a **histogram**.

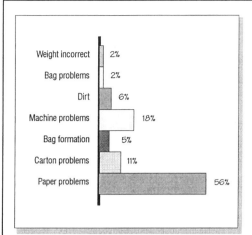

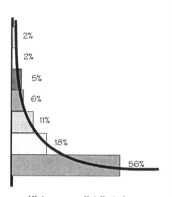

Histogram redistributed to show Pareto effect

The operators found 7 areas where tea was lost. They checked on the chain of processes from the moment the tea came in by lorry, through when it was weighed to when it was poured into the hopper. Then they checked on the operation of bagging the tea where the two giant paper rolls were brought together around measured quantities of tea, perforated, then inserted into cartons.

By far the biggest loss was due to problems involved in changing over the paper rolls when one or the other of them ran out. Notice how the proportion of problems was distributed according to Pareto's law - in this case 74% of the waste came from 28% of the identified problem areas.

So the team decided that, if the company ordered larger rolls and underwent some minor machine changes to accommodate them, there would be fewer changeovers, and so less wastage.

There were two other bonuses. Here's one:

And the other was this:

The increased attention paid to the changeover by the operators resulted in greater care and therefore even less spillage. The result was that in six months the spillage had been reduced by 75%, well over Liz's target.
The intelligent use of data collection tools had identified the problem and suggested the remedy.

Once a process can be controlled statistically, variations in quality can be regulated and reduced

1   For outputs to meet the requirements of the customer, inputs must be defined, monitored and controlled.

2   Recorded data help organizations to track down inefficiency and increase profitability. The tool kits below aid this process.

3   *Process flowcharting* shows in pictorial form what is done.

4   *Check sheets/tally charts* assist in the collection of data to discover how often it is done.

5   *Histograms* present the data to show what the overall variations look like.

6   *Scatter diagrams* show the relationship between factors.

7   *Pareto analysis* leads to the discovery of the big problems (typically 80% of the problems arise from 20% of the causes).

8   *Cause and effect analysis* and *brainstorming* give a "fish-bone" diagram of what causes the problems.

9   *Force-field analysis* is used to discover what will obstruct or help the change or solution.

10   The *emphasis curve* is a simple technique to agree on what are the most important factors.

11   *Control charts* are fundamental to understanding which variations to control, and how.

12   The appropriate use of some or all of these techniques, separately or together, can help people solve problems and make process improvements.

No one is likely to argue with the proposition that teamwork is important to the success of an organization

Teamwork

Yet, teamwork is not always easy to achieve...

WE MUST MAKE WHAT WE CAN SELL

WE MUST SELL WHAT WE CAN MAKE

...with departments or individuals having different corporate objectives...

...to say nothing about personal or departmental rivalries.

And, although we may have an idealized notion of teamwork permeating the whole organization...

ALL TOGETHER, BROTHERS

Good sport, Alice, ha ha

HEAVE, TEAM

...this need not necessarily be true.

SO...

A The only efficient way to tackle process improvement or complex problems is through teamwork. It allows individuals and organizations to grow.

Employees will not effect continual improvement without commitment from the top, a 'quality' climate and an effective mechanism for capturing individual contributions.

Teamwork for quality improvement is driven by a strategy, needs a structure and must be implemented thoughtfully and effectively.

Teamwork allows everyone in the organization to become part of the process of change

On the ted final assembly line...

....there is a team of people who have worked together for some time.

All of them are engaged in the assembly of teddy bear components, but the job of the last one in the line...

...is the inspection of the finished bears. But he has another job..

When a fault is discovered by any member of the team...

...the alarm is raised...

...and work on the line stops immediately as the cause of the holdup is identified by all the team.The team analyses the holdup, repairs the defect if possible, tags the defective part, and the line gets working again.

Meanwhile the team member with the task of checking the final product does his other job.

He would do this by using a simple device - a "chumbo chart"- columns of clear plastic tubes into which he drops white marbles. Each column corresponds to a stage in the process or type of fault. The chumbo chart is a continuous display of main source of faults or process problems.

If for any reason the fault gets past the inspect and repair man the next operator down the line would put a red tag on the defective item and a red marble would be put into his chumbo chart, and serious questions asked about the process.

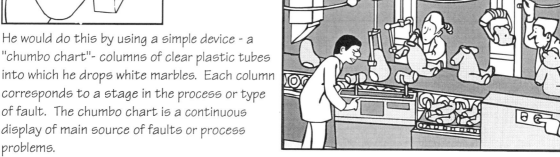

The team has a rota of jobs which means that a different member has the inspection job every day. So the responsibility for making the system work is shared equally amongst all the members of the team, and there is no element of blame attached to failures or stoppages.

Successful teamwork means a no-blame environment where every member is responsible for improvement in the job they do

But, as much as organizations need teams, and teams need individuals, most individuals need to be part of a group

A social psychologist called Maslow discovered that people's needs varied according to the circumstances they found themselves in...

At the very lowest level of Maslow's hierarchy of needs is the sheer need for survival – the physiological imperatives of satisfying hunger, thirst and the need for sleep.

Then comes the need to feel secure and protected from danger

When people are confident of not going hungry or becoming homeless, their next need is to be accepted by other people. They develop a need for love, friendship and membership of a group.

...and when membership of the group is assured, they need the recognition of the group and the enhancement of their self-respect.

SHAREHOLDERS' REPORT

Finally comes the need for personal growth and accomplishment in the individual's own terms.

MY LIFE
A Novel

So as the individual climbs through the stages of Maslow's hierarchy the need to form groups becomes stronger...

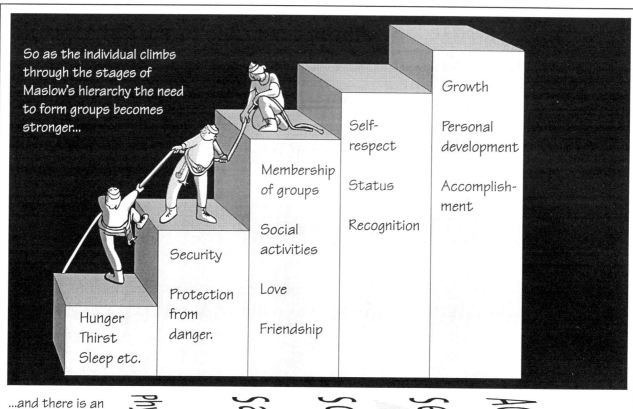

Growth

Personal development

Accomplish-ment

Self-respect

Status

Recognition

Membership of groups

Social activities

Love

Friendship

Security

Protection from danger.

Hunger
Thirst
Sleep etc.

Physiological

Safety

Social

Self-esteem

Actualization

...and there is an in-built instinct for self improvement. But there is one thing to remember.

Just as it is possible to climb from one level to another...

...it is equally possible to go down...

I can give you a receipt for tax purposes

Down-sized accountant

...as many people have discovered.

So people have a natural tendency to form groups, for their own sense of self-preservation. These groups can become creative teams when they are given a focus, or a common cause, and when they can produce an acceptable leader.

But what is required of a leader?

The main block, with the old pavilion (arrowed)

In a certain company, the sports club wanted to replace its decrepit pavilion. The management agreed to provide enough funds for the materials. It was agreed that the sports club members would build it

The president of the sports club, Glenda Bishop, was put in charge of the project.

Goodie! I'm FULL of ideas!

Glenda looked for volunteers for the scheme by placing a notice on the sports club notice board...

and she soon had four members keen to help. But from the beginning, Glenda saw herself as boss...

You, Tommy, can phone around for materials, and, Kitty...

Hey!

...and the reaction was immediate.

You're not the only one with ideas! Besides, if anyone should be phoning around it's me - I run Procurement!

...and Glenda was faced with rebellion in the first half hour of the project.

What went wrong?

The guru helps Glenda...

If it's not handled carefully, any attempt to form a team can end in tears, and not just for the leader either.

The basic thing to remember about teamwork is that it has three elements, or needs. and they ALL have to be satisfied.

The three elements of teamwork

Task needs

For a team to succeed, it needs to have a common purpose, i.e. a task agreed by all members of the team.

Team needs

For a team to succeed, every member must pull in the same direction, to be held together and to agree on ways and means.

Individual needs

For a team to succeed, individual members must know and be happy with what contribution they have to make

The functions of a leader

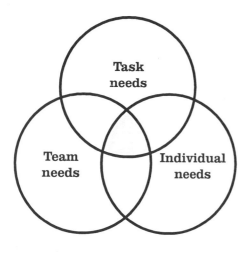

The team leader has responsibilities in each of the three areas, and must co-ordinate them. So the main job of the team leader lies within the overlap of the three elements of teamwork.

Within each area the leader's functions are:

Task: Defining the task, making a plan, allocating work and resources, controlling the quality and tempo of work, checking team performance, adjusting the plan.

Team: Setting standards, maintaining discipline, building team spirit, encouraging and motivating, appoining sub-leaders, ensuring communication within the team.

Individual Attending to personal problems, praising individuals, giving status, recognizing and using individual abilities, training the individual.

With grateful acknowledgements to J. Adair, *Effective Teambuilding,* Pan Books

A leader's checklist for each of the three elements

The following checklist should help the team leader to measure the progress of the team against the required functions of fulfilling the task, maintaining the team and allowing the individuals within the team to develop.

Task

1	Are the targets clearly set out?
2	Are there clear standards of performance?
3	Are available resources defined?
4	Are responsibilities clear?
5	Are resources fully utilized?
6	Are targets/standards being defined?
7	Is a systematic approach being used?

Team

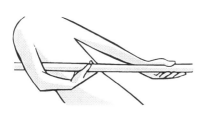

1	Is there a common sense of purpose?
2	Is there a supportive climate?
3	Is the unit growing and developing?
4	Is there a sense of corporate achievement?
5	Is there a common identity?
6	Does the team know and respond to the leader's vision?

Individual

1	Is each individual accepted by the leader/team?
2	Is each individual made part of the team by the leader/team?
3	Is each individual able to contribute?
4	Does each individual know what is expected in relation to the task and by the team?
5	Does each individual feel a part of the team?
6	Does each individual feel valued by the team?
7	Is there evidence of individual growth?

In this case, reconciliation led to a better relationship between the leader and the team - and between individual team members. Crucially, it also led to the establishment of TRUST

Trust

is the cornerstone of successful team relationships.

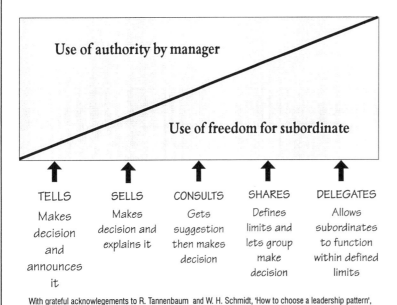

Use of authority by manager

Use of freedom for subordinate

TELLS	SELLS	CONSULTS	SHARES	DELEGATES
Makes decision and announces it	Makes decision and explains it	Gets suggestion then makes decision	Defines limits and lets group make decision	Allows subordinates to function within defined limits

With grateful acknowlegements to R. Tannenbaum and W. H. Schmidt, 'How to choose a leadership pattern', *Harvard Business Review*, May-June, 1973

Continuum of leadership behaviour

The developing relationship between the team leader and the group can be represented by this graph which shows that, as the need for the use of authority by the manager decreases, the area of freedom for the subordinates, and therefore the amount of decision-making within the team increases.

Mature and successful teams will refer to the leader for task direction in progressively fewer cases as the task nears completion.

Stages of leadership

At the beginning of the team's life together the leader will be very directive, giving clear instructions to meet agreed goals.

As the team becomes more experienced and successful the leader will adopt a more coaching approach.

The next stage will be to allow more initiative by the members of the team,

with some help from the leader,

Ultimately, the leader adopts a delegating style, and is able to take a role as team member.

Delegation can only take place with developed 'followers'.

The way leadership styles change through various stages of team development is illustrated by this diagram showing how the leader's supportive and directive behaviour continues to change as the team activity progresses.

Beginning at the bottom right hand corner, the diagram shows that a leader's style is likely to be highly directive and not very supportive (area S1) and proceed through highly supportive/directive (S2) and highly supportive/less directive (S3) phases to the point where he or she needs to supply only low support and direction (S4).

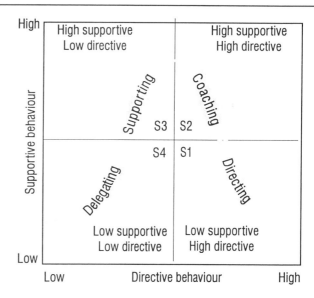

With grateful acknowledgements to K. Blanchard and P Hersey, *Management of Organizational Behaviour: Utilizing Human Resources*, Prentice-Hall

Some managements believe that it is possible to put a team together at S1 and expect it to proceed immediately to its mature state at S4, in effect tunneling through the painful hill of experience which climbs through S2 and S3. It cannot be done.

Forming, storming, norming and performing

Four stages in the evolution of a team

1 Forming

- Feelings covered up
- People conform to established lines
- No account taken of others' values and views
- No shared understanding of task

2 Storming

- More risky, personal issues are opened up
- The team becomes more inward-looking
- There is more concern for the values, problems of others in team

3 Norming

- Confidence, trust within the team
- More systematic approach
- More value given to other
- Clarification of objectives
- All options considered
- Detailed plans prepared
- Progress review for improvement

4 Performing

- Flexibility
- Leadership decided by situation, not protocol
- Everyone's energies utilized
- Basic principles and social aspect of the organization's decisions considered

With grateful acknowledgements to B. W. Tuckman and M.A. Jensen, 'Stages in small group development revisited', *Group Organizational Studies*, 2 (4), 1977

Life cycle of a team

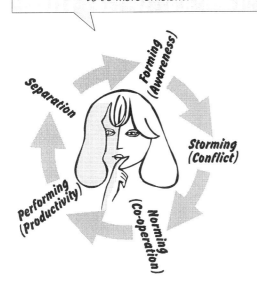

If the team continues with another project, the forming, storming and norming phases are shorter and the team is likely to be more efficient.

Forming (Awareness)

Separation

Storming (Conflict)

Performing (Productivity)

Norming (Co-operation)

Glenda Bishop now knows about the evolution and development of her relationship with her team, from its first stage of forming, through the conflict of the storming phase, norming and finally producing results - performing. After that the team may be disbanded, or possibly go on to use their team skills on another project. During her time as leader Glenda will become aware of some questions:

■ How is leadership exercised in the team?

■ How is decision-making accomplished?

■ How are team resources utilized?

■ How are members integrated into the team?

The answers to these questions are not easy, and are different for every team and for every set of circumstances, but they are fundamental questions which have to be addressed by the team leader if he or she is to be successful.

Attributes of a successful team

No group can be effective without knowing what they want to achieve. People are more committed when they identify with goals. Goals need to be agreed by members.

Team members need to be able to express opinions without fear of ridicule. Cut-throat atmosphere means loss of creativity.

Openness breeds trust - the essential ingredient for successful teamwork.

Individuals listen to others and ideas are built on them. Co-operation causes high morale and enables contributions to be made from the team's pool of experience. Conflicts are a necessary part - the effective team works through issues of conflict to achieve objectives. Conflict prevents teams becoming complacent and lazy.

Effective decision-making means quick information-gathering, discussion, decisions.

Leadership must be appropriate to the team make-up. No one style is correct. Leadership role changes as the team develops.

Effective teams review their performance, use of conflict etc to perform better. Feedback is encouraged - from outside if necessary.

The best team relationships allow room for individual development.

The team is a unit, and the members are components of that unit...

Teamwork is a process

Like other forms of corporate activity, problem-solving through teamwork is a process, with the team providing an input or inputs, with the expectation of an output or outputs.

For this reason, all energies and efforts have to be directed toward the team input. This means keeping tight control of the self-orientation of individual team members.

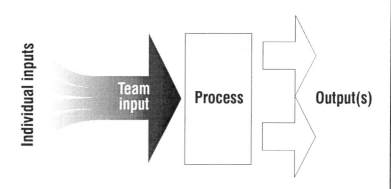

Individual inputs → **Team input** → **Process** → **Output(s)**

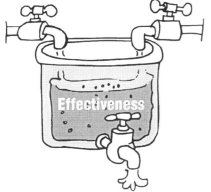

Task fulfilment   **Team maintenance**

Effectiveness

Self-orientation

To produce an effective output, teams must have:

* **high task fulfilment (+);**
* **high team maintenance (+);**
* **low self orientation (-)**

So, while high task fulfilment and high team maintenance factors increase the effectiveness of the team, individuals in the group who have an excessive need to assert their personal preferences at the expense of teamwork will reduce it.

Factors which affect a team's performance

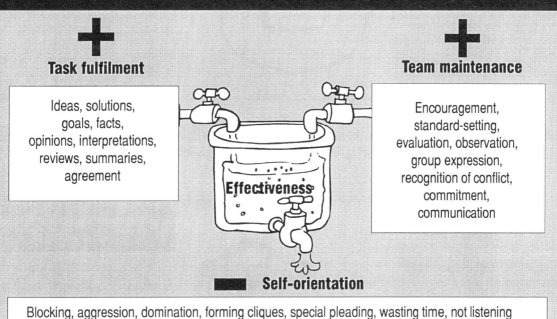

+

Task fulfilment

Ideas, solutions, goals, facts, opinions, interpretations, reviews, summaries, agreement

+

Team maintenance

Encouragement, standard-setting, evaluation, observation, group expression, recognition of conflict, commitment, communication

Effectiveness

Self-orientation

Blocking, aggression, domination, forming cliques, special pleading, wasting time, not listening

The task of the team leader is to encourage high task-fulfilment and team maintenance while discouraging indivduals' tendencies toward self-orientation.

Characteristics of team members

Who do you think you are, with your big ideas?

Who indeed?

Who am I?

What am I?

Because individuals vary so enormously, it is useful for a team leader - in fact all team members - to be aware of the **preferences** which determine their thoughts, feelings and actions, and those of their fellow members.

A reliable way of getting a personal profile is to use the Myers-Briggs Type Indicator (MBTI). This is based on an individual's preferences on four scales:

giving and receiving energy, gathering information, making decisions and handling the outer world.

The four MBTI preference scales represent two opposite preferences in each of the four scales

Extroversion-Introversion
(how we prefer to give/receive energy or focus our attention)

Sensing - Intuition
(how we prefer to gather information)

Thinking - Feeling
(how we prefer to make decisions)

Judgement - Perception
(how we prefer to handle the outer world)

People generally have a **preference** on each of these scales. For example, people with a preference for **thinking** tend to make decisions based on impersonal logical analysis.

Preference is like left- or right-handedness: neither is right or wrong, but we feel natural, easy and efficient when using the preferred hand. The MBTI is therefore not a test of ability, but an index of the way people go about tackling problems.

But which is my **true** preference?

Sometimes I have to act the **extrovert**...

But then...

...afterwards I just want to hide and recharge my batteries...

...because his true preference is introversion, to get energy from the inner world of thoughts and ideas.

Identifying the characteristics of individuals

There are eight possible preferences, i.e. two opposites for each of the four scales.

Extroversion
(which we will call **E**) versus
Introversion (I)

Sensing (S)
versus
Intuition (N)

Thinking (T)
versus
Feeling (F)

Judgement (J)
versus
Perception (P)

E/I	S/N
T/F	J/P

Hi there, welcome to the team. I'm ENFP

Pleased to meet you, old chap, I'm ISTJ

Team Get-together

An individual's type is the combination and interaction of his or her four preferences and can be assessed initially by completion of a simple, non-threatening questionnaire.

This will result in a four-letter code.

The code is a sequence of a choice of one of the two possibilities in each of the four preferences.

Some combinations of characteristics

is an ESTJ and prefers extroversion (E), i.e. is energized by the outer world and things. He prefers to gather information by sensing (S), prefers to make decisions by by thinking (T) and has a judging attitude (J) towards the world (i.e. prefers to make decisions rather than go on collecting information).

is an INFP, and prefers introversion, focussing on the inner world of thoughts and ideas. He prefers intuition for perceiving the world, feelings or values for making decisions, and likes to maintain a perceiving attitude towards the outside world (i.e. gathering information rather than making a decision).

is an ISFJ. She prefers introversion and likes to work quietly without interruption. A practical type, she prefers to concentrate on the reality of a situation. She enjoys dealing harmoniously with people and likes to be organized.

is an ESFP, who prefers communication to introspection, is good at working with a large number of facts, especially about people; is sympathetic and tactful to others, and is always open to new experiences.

A key function of the team leader is to use the preferences of the members productively. Here the leader allows an extrovert to bully an introvert, with counter-productive results...

So, how do we all feel about it? Colin?

Er...

Well obviously we go ahead on it

Stands to reason, eh, Col?

I mean, it's common sense isn't it?

Um..

All agreed then?

Yes, Gloria

OK, I declare this meeting closed

I mean..

Anything to add, Colin?

Gulp...

Cheerio then

Short term we'd save money, but over the period of the project it would mean massive redundancies so I am **completely against it!**

SLAM!

The MBTI as an aid to teamwork

There are 16 combinations of the preference scales

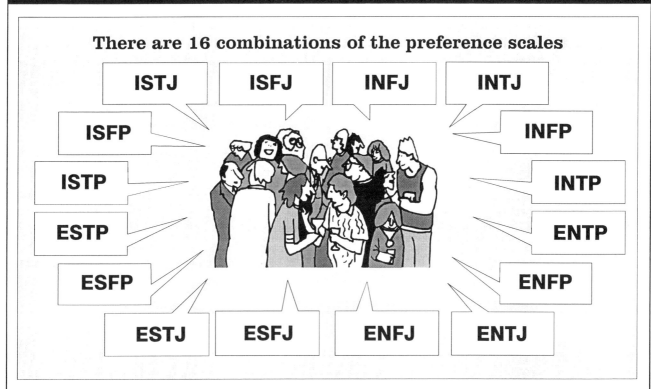

ISTJ ISFJ INFJ INTJ ISFP INFP ISTP INTP ESTP ENTP ESFP ENFP ESTJ ESFJ ENFJ ENTJ

The team leader will find that intelligent use of people's preferences can help form teams which will perform very successfully.

He or she must know that:

Extroverts prefer action and the outer world.

Introverts prefer ideas and the inner world.

Sensing-thinking types are interested in facts and analysis and apply their findings impersonally.

Sensing-feeling types, although interested in facts, relate their analyses to themselves and others.

Intuition-thinking preferences result in an interest in possibilities and have theoretical, technical or executive abilities.

Intuition-feeling combinations, although interested in possibilities, prefer tackling new projects, unsolved problems.

Judging types are decisive and like planning and control.

Perceivers are flexible, live spontaneously, and understand and adapt readily.

If individuals in a team are prepared to share their MBTI preferences with the others, this will increase understanding and lead to better team performance.

> The assistance of a skilled MBTI practitioner is essential in the initial stages of this work

Problem-solving using type preferences

Problem-solving should follow a logical path: get the **facts**, examine the **possibilities**, assess the **consequences** of any resulting action, and finally gauge the impact the implementation of the plan may have on the people affected. On the right is that path showing the stages at which certain preferences are likely to play a strong part. A well-integrated team is able to use the pre-eminent preferences of individuals intelligently to get the best results at each of the four stages.

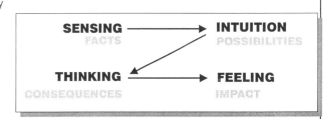

SENSING → INTUITION
FACTS POSSIBILITIES

THINKING → FEELING
CONSEQUENCES IMPACT

However, in practice this problem-solving model can be difficult to use by individuals. Because of the bias given by an individual's preferences, not every stage will be given equal attention. So a sensing-thinking type might concentrate on the first and third stages and neglect the other two, or, diagrammatically:

For a team to use the **MBTI** and work together successfully they should:

1 be made <u>aware</u> of it;

2 <u>accept</u> the principles as valid;

3 <u>adopt</u> them for themselves in order to

4 <u>adapt</u> their behaviour accordingly.

This will lead to

5 team <u>action.</u>

Implementing teamwork for quality improvement

The MBTI problem-solving zig-zag path is valuable when it is applied to particular tasks by teams which can use the strengths of individual members. It translates into a process by the addition of some problem-solving steps:

D (define the problem),
R (review the information),
I (investigate the problem),
V (verify the solution) and
E (execute the change).

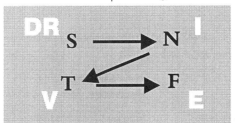

The steps relate to the original path like this

In our final chapter we will look at how the various elements of the TQM model are utilized to solve the problem a company had of delivering its goods on time.

1   Teamwork is vitally important to the success of any organization, yet not always easy to achieve.

2   Process improvement or solving complex problems should be tackled through teamwork, which allows individuals and organizations to grow.

3   Teamwork for quality improvement needs a strategy, structure and careful implementation.

4   Maslow's hierarchy of needs - physiological, safety, social, self-esteem and actualization - leads to a natural tendency to form groups.

5   Groups can become teams when they understand the needs of the task, the individuals and the team - Adair's model for action-centred leadership.

6   Leadership may be considered to move through a continuum of *tells - sells - consults - delegates* (Tannenbaum & Schmidt); or as situational: *directing - coaching - supporting - delegating* (Blanchard).

7   Team development typically passes through the stages of *forming, storming, norming and performing* (Tuckman).

8   To produce effective output, teams must have **high** task fulfilment and maintenance with **low** self-orientation.

9   The Myers-Briggs Type Indicator (MBTI) is useful in helping individuals in teams to understand and value each other. This has four preference scales:

        E-I, S-N, T-F, J-P

giving sixteen possible personality types.

10   The MBTI S-N and T-F scales may be matched to a problem-solving framework, DRIVE (define, review, investigate, verify and execute) which helps teams achieve their objectives.

9 *Implementation*

To see the DRIVE model in action, let's take a typical problem situation and see how the process works when DRIVE is applied to it. Here's the problem:

One morning at Didgets...

Didgets? This is Fidget of Fidgets. Send me some widgets...NOW!

Right, Mr Fidget. I'll get them off to you today

R.LIDGET

How are we off for widgets, Helen? I need 5000

Widgets is hardware. I'm software, Mr L.

Hardware? They've moved across the road, I think

No, we're Large Items. You want Harry, but he's off with 'flu

Oh well, I tried...

Tomorrow is another day

Later that week

How are you, Harry? You look rough

Not so hot, but I'll cope

But Harry's illness makes him forget the order for two days...

...and even then its progress is not without mishap...

...including the wrong address on the order

Sid, I'm telling you there's no Bridget Lane in this town

With the inevitable result...

Is that Lidget of Didgets? Where's me widgets? You can keep your order. I'll go to Midgets!

We've lost a valuable customer because several things went wrong, any one of which could have lost us the order

Yes, sorry, chief, we all feel bad about that

I accept my part of the blame, but we must rethink our whole attitude to supplying our customers, and to that end...

...I have asked the internationally famed guru to come and advise us

Thanks, Mr Didget, and hello everybody!

He's already part way to changing the culture and getting commitment

I'm going to ask you first to form a team. To help us in this we'll start by using the Myers-Briggs Type Indicator to determine your character preferences

Then I will want the team to look at all the processes involved in your business activities using the problem-solving tools I'll tell you about....

OK, the MBTI has given us an idea of the characteristics of the team, and when we've sorted out who's team leader we can look at the problems in turn

Chris, the Sales Manager, was an ENTJ and a natural choice for team leader. His social skills made each member of the team feel important and he will help them reach a decision.

Helen, an order processing clerk, ISFP, is a sympathetic, perceptive and tactful introvert who will be able to persuade people to implement change without friction.

Harry, a marketing assistant, ESFP, is outgoing and energetic. His feeling for nuances allows him to extract vital information from the most obstinate employees.

Douglas, stores supervisor, INTJ, is quiet and well organized. He is challenging and skilful at implementing team decisions.

How can he be so confident without a system? Does he know who his supplier within the company is?

Helen is the wrong person to ask. He should know who his supplier is - they should have had the relevant process sorted out long ago.

There's a strong case for improved communications within the firm

A system is needed which doesn't rely on individuals, however valuable they are.

Waiting for Harry has set the order back. We have to create robust processes and record data so that Harry's flu doesn't have such an impact.

We must not forget there there are many processes to DRIVE (define, review, investigate, verify and execute) and IMPROVE.

First, we identify where the main problem lies...

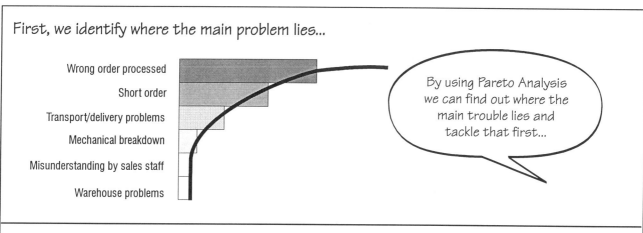

Wrong order processed
Short order
Transport/delivery problems
Mechanical breakdown
Misunderstanding by sales staff
Warehouse problems

By using Pareto Analysis we can find out where the main trouble lies and tackle that first...

Then we set about solving it.

...and by installing proper **systems** by means of analysing the flow of work, writing down what **should** happen, we can improve each

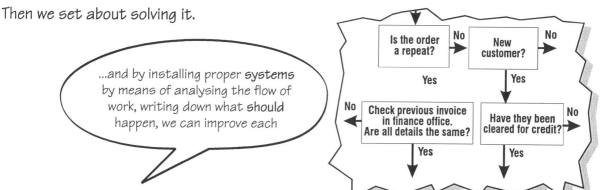

Is the order a repeat? — No → New customer? — No →

Yes ↓ (Is the order a repeat) ... Yes ↓ (New customer)

No ← Check previous invoice in finance office. Are all details the same?

Have they been cleared for credit? — No →

Yes ↓ (Check previous invoice)

Yes ↓ (Have they been cleared for credit)

The DRIVE model formed the basis of the team's activities

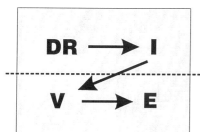

$$DR \rightarrow I$$
$$V \rightarrow E$$

D- define the problem; R - review the information; I - investigate the problem. The first half of the DRIVE model was used by the team to accumulate the information required to make the necessary changes, and used the sensing and intuition preferences of the team.

V- verify the solution; E - execute the change. The second half of the model called on the thinking and feeling skills of the team members to implement the change with the least friction.

The result...

We gradually evolved an operation which was virtually error-free, and we were able to delight the customer by providing a better service than he expected.

The DRIVE model gives the problem-solving steps from identifying problem areas to providing then implementing a solution.

The next page shows how the DRIVE model was put into practice in this case.

FANTASTIC! How did you know I was just going to order more widgets? You know more than my stock control department!

Our sales office saw that your previous ordering pattern indicated that you were likely to be low, so we brought some just in case, Mr Fidget

The implementation process in action

In solving the problem of delivery on time, the team used the DRIVE model like this:

DEFINE:
Is this a problem area?
What is the real problem?
Does it apply to all customers?
Can we successfully tackle this problem?
How can we define and prioritize our tasks?

REVIEW:
From existing records, time elapsed from receiving order to delivery is averaged out.
Improved schedule is agreed by the team.

INVESTIGATE:
The team construct flowcharts of the various stages of receipt of order to delivery.
Grey areas like orders for new products are noted (not enough information).
Flowpath is modified to include technical briefing of sales staff for new products.
Recommendations made to cover staff illnesses by overlapping responsibilities.
Physical handling of goods in Despatch reviewed.
A check-sheet is designed.

VERIFY:
The modified process is implemented.
The check sheet is used to gather data.
Turn-around time is monitored.
Devices for warning of lateness and mishaps are checked.

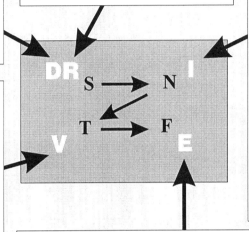

EXECUTE:
Data is presented to a meeting of the sales, technical and despatch departments.
Changed procedures are agreed, documented and circulated.
A documented procedure for chasing orders, with a 'no blame' warning when things go wrong, is implemented.
The new system is monitored.

This method brings together the TQM process-based philosophy, with its continuous assessment of performance, with team planning in which each member can identify activities which relate to his or her individual skills. It is an indispensable part of the whole TQM model. It links teams, systems and tools to produce a strong and enduring quality-based company-wide method of ensuring continuous improvement.

The TQM model in action

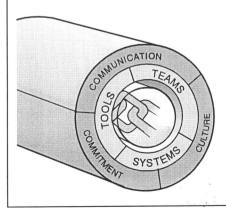

The company's problems were identified in the first place by looking at its PROCESSES. A TEAM of people analysed the situation using problem-solving TOOLS. Once a new process was designed the SYSTEM was re-documented and the relevant people are trained accordingly.

The CULTURE of the company became one of tracing problems and creating positive prevention systems, rather than one of blame. There was a COMMITMENT to improving processes, and good COMMUNICATIONS delivered the existence - and the solution - of the problems to the workforce.

The TQM model is applicable to all organizations in this way.

924080